PHYSICAL AND NON-PHYSICAL METHODS OF SOLVING CRYSTAL STRUCTURES

PHYSICAL AND NON-PHYSICAL METHODS OF SOLVING CRYSTAL STRUCTURES

MICHAEL WOOLFSON AND FAN HAI-FU

CAMBRIDGE
UNIVERSITY PRESS

Published by the Press Syndicate of the University of Cambridge
The Pitt Building, Trumpington Street, Cambridge CB2 1RP
40 West 20th Street, New York, NY 10011-4211, USA
10 Stamford Road, Oakleigh, Melbourne 3166, Australia

First published 1995

Printed in Great Britain at the University Press, Cambridge

A catalogue record for this book is available from the British Library

Library of Congress cataloguing in publication data
Woolfson, M. M.
Physical and non-physical methods of solving crystal structures /
Michael Woolfson and Fan Hai-fu.
p.   cm.
ISBN 0 521 41299 4
1. X-ray crystallography – Technique.   I. Fan, Hai-fu.   II. Title.
QD945.W59   1995
548′.83–dc20   94-8254 CIP

ISBN 0 521 41299 4 hardback

Contents

Preface

When von Laue and his assistants produced their first smudgy X-ray diffraction photographs in Munich in 1912 they could not have known of the developments that would follow and the impact that these would have on such a wide range of science. Structural crystallography, the ability to find the arrangement of atoms inside crystals, has advanced over the years both theoretically and experimentally. Technical advances, such as the development of computers both for control of instruments and for complex calculations, and also the advent of high power synchrotron X-ray sources have all played their part.

In this book we bring together all the methods that have been and are being used to solve crystal structures. We broadly divide these methods into two main classes, *non-physical* and *physical* methods. In the first category we place those methods that depend on a single set of diffraction data produced by the normal Thomson scattering from the individual atoms. The Patterson methods and direct methods described in chapters 2 and 3 respectively are non-physical methods. In chapter 4 the basic principles are explained for two physical methods – isomorphous replacement, which combines the data from two or more related compounds to obtain phase information, and anomalous scattering, which uses data at wavelengths for which some of the atoms scatter anomalously, i.e. with an amplitude and phase differing from that given by the Thomson process. In chapter 5 the method of isomorphous replacement is explored in much greater depth and in chapter 6 the same is done for anomalous scattering. It will be seen that some of the most effective ways of using *physical* data are in conjunction with *non-physical* methods, in particular direct methods, although Patterson function ideas also come into one successful technique for using anomalous scattering data.

The outcome of physical approaches, particularly when applied to macromolecular structures, is that phases are found only for low resolution data although data at higher resolution might be available. This requires the techniques of phase extension and refinement which are described in chapter 7. Here we bring in some material slightly off the beaten track of conventional X-ray crystallography concerned with the solution of crystal structures by a combination of electron microscopy and diffraction and also the study of crystals with pseudosymmetry of various kinds.

In chapter 8, the final chapter, we describe what may be considered the most physical of all methods, which exploits the dynamical scattering process to acquire phase information. One of the helpful features of X-ray diffraction is that dynamical scattering effects are weak and this makes the connection between the observed data and the electron density in the crystal much more direct. It is therefore paradoxical that by doing very refined and accurate experiments to detect dynamical scattering effects one can go a long way towards solving the phase problem, the basic problem of structural X-ray crystallography.

We hope that this book will be found useful by working crystallographers and, in particular, those, such as graduate students, who are being introduced to the problems of solving crystal structures for the first time. For our readers who are at the beginning of their crystallographic studies we have given in chapter 1 an introduction to the field, sufficient to give comprehension to what follows without the pretence of being a complete treatment. Indeed we would not pretend that within the confines of a book of modest size we have covered exhaustively all the topics with which we have dealt. If we can give understanding and point the way to further reading and study where required then we shall have achieved our purpose.

1

The basics of X-ray diffraction theory

1.1 Forming an image

The process of forming an optical image is one that is very well understood and frequently occurs. In the very act of seeing what is on this page the reader is forming a retinal image of its contents which is then conveyed to the brain in the form of electrical impulses for the complex task of interpretation and comprehension. What happens in the visual cortex is poorly understood but the formation of the retinal image via the lens of the eye is straightforward and can be followed by reference to fig. 1.1. The first stage in image formation is to direct towards the object some radiation (light in this case), part of which is scattered so that each point of the object becomes a secondary source of radiation which leaves in all directions. If we look in detail at what happens at a point of the object (fig. 1.1(a)) we see that the radiation going off in different directions is not only coherent, because it derives from the same point source, but is also all in phase. Next the scattered radiation strikes a lens (fig. 1.1(b)). Because the speed of light in the lens material is different from that in air, rays travelling by different paths to the image point have the same optical path length and so undergo constructive interference there. The amplitude, and hence intensity, of each image point is proportional to that of the corresponding object point and consequently a true image is formed.

1.2 Scattering from a periodic object

If we looked at the scattered light itself by putting a screen between the object and the lens then we would find that it consisted of a rather formless arrangement of lighter and darker patches with no obvious correlation with the object. However, there are some types of object with which the pattern of scattered radiation does seem to be highly correlated. For example,

1

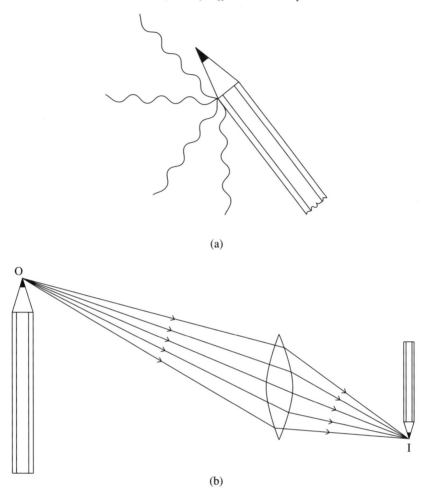

(a)

(b)

Fig. 1.1(a) Waves scattered by a particular point of the object are all in phase. (b) Because light travels more slowly in glass than in air all optical paths from O to I have the same duration.

consider an object consisting of a regular linear array of similar small apertures as shown in fig. 1.2. The scattering pattern from this will be in the form of a series of well defined and separated beams. If the oncoming radiation falls normally onto the object (fig. 1.2(a)) then the condition for a beam to be formed is that on the scattered wavefront the radiation scattered from neighbouring apertures should be in phase, i.e.

$$a \cos (\alpha) = h\lambda, \tag{1.1}$$

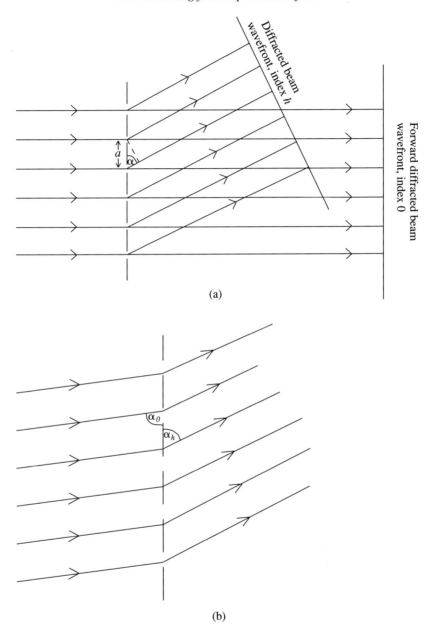

Fig. 1.2(a) Formation of diffracted beams of index 0 and *h* with a normal incident beam. (b) Diffraction for incident beam in a general direction.

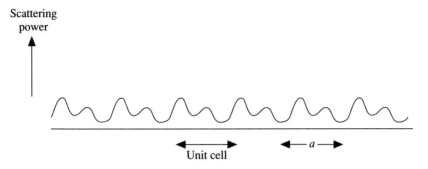

Fig. 1.3 Representation of scattering power as a function of position for a portion of a one-dimensional periodic scattering object (diffraction grating).

where a is the spacing of the apertures, α the angle shown in fig. 1.2(a), λ the wavelength of the radiation and h some integer. In fact such a scattering object would normally be referred to as a *diffraction grating* and such a scattering process is known as *diffraction*. We also refer to the beam defined by equation (1.1) as having an *index h*. If the oncoming beam fell on the grating at some general angle α_0 (fig. 1.2(b)) then the beam of index h would lie in a direction such that

$$a\left[\cos(\alpha_h) - \cos(\alpha_0)\right] = h\lambda. \tag{1.2}$$

The directions of the diffracted beams indicate the periodic nature of the diffracting object but if the similarly diffracting apertures were structured in some way then information about that structure would be contained in the relative amplitudes, or intensities, of the beams. The amplitude of the *diffracted* beam would be proportional to the amplitude of scattering of a single scattering unit at the appropriate angle. Actually, although we have been using the imagery of scattering apertures separated from each other by non-scattering or opaque regions, this is an unnecessary restriction. The analysis given above can apply equally well to periodic diffracting objects, which diffract everywhere; in fig. 1.3 there is shown a representation of such an object consisting of contiguous one-dimensional *unit cells*.

It is a straightforward matter to extend the above description to the case of scattering from a three-dimensional periodic object. In this case, if the periodicity of the object is described by the three displacement vectors **a**, **b** and **c**, the equations for the constructive interference of scattering from neighbouring cells are

$$a\left[\cos(\alpha_h) - \cos(\alpha_0)\right] = h\lambda,$$
$$b\left[\cos(\beta_k) - \cos(\beta_0)\right] = k\lambda,$$

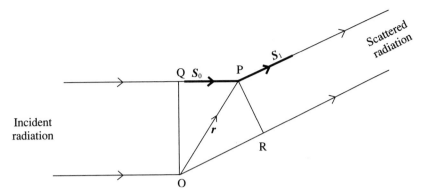

Fig. 1.4 Scattering from the point P. The incident wavefront is represented by OQ and the scattered wavefront by PR.

$$c\left[\cos(\gamma_l) - \cos(\gamma_0)\right] = l\lambda, \qquad (1.3)$$

where the incident beam makes angles α_0, β_0 and γ_0 with the directions defined by $\mathbf{a}$, $\mathbf{b}$ and $\mathbf{c}$ and the scattered beam of indices (h, k, l) makes angles α_h, β_k and γ_l with those directions. There is, however, an important difference between the one-dimensional and three-dimensional cases. For any direction of incident beam in the one-dimensional case, as long as h is not too large, some value of α_h will be found to satisfy equation (1.2) whereas for an arbitrary incident beam defined by α_0, β_0 and γ_0 there may be no diffraction direction giving values of α_h, β_k and γ_l that simultaneously satisfy equations (1.3). Nevertheless, for a range of values of h, k and l, directions can be found for an incident beam, which will permit solution of equations (1.3) (see section 1.5.1) and that is all that immediately concerns us here.

As was previously stated for the one-dimensional case, when a scattered beam with indices (h, k, l) is formed it will have an intensity proportional to the intensity of scattering from the contents of a single three-dimensional unit cell and we shall now consider what this would be in particular cases.

1.3 Scattering from a non-periodic object

1.3.1 A single point scatterer

Consider radiation in the form of a plane wave falling on a point scatterer, P, with vector position $\mathbf{r}$ relative to some origin, O, as shown in fig. 1.4. The direction of the incident radiation is defined by the unit vector $\mathbf{S}_0$ and we

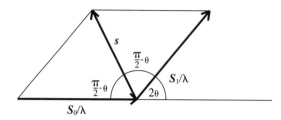

Fig. 1.5 The relationship of the scattering vector **s** and the unit vectors $\mathbf{S}_0$ and $\mathbf{S}_1$.

take the scattered radiation in a direction defined by the unit vector $\mathbf{S}_1$. It is clear that the scattering from P is ahead of that which we would get from a scatterer at O by a path difference

$$OR - PQ = \mathbf{r} \cdot \mathbf{S}_1 - \mathbf{r} \cdot \mathbf{S}_0 = \mathbf{r} \cdot (\mathbf{S}_1 - \mathbf{S}_0).$$

This path difference is equivalent to a phase difference of

$$\phi_P = 2\pi \mathbf{r} \cdot (\mathbf{S}_1 - \mathbf{S}_0)/\lambda = 2\pi \mathbf{r} \cdot \mathbf{s}, \tag{1.4}$$

where

$$\mathbf{s} = (\mathbf{S}_1 - \mathbf{S}_0)/\lambda. \tag{1.5}$$

In scattering theory **s** is an important quantity, which we call the *scattering vector*. In fig. 1.5 we show its relationship to the unit vectors $\mathbf{S}_0$ and $\mathbf{S}_1$. For historical reasons, connected with an early description of X-ray diffraction by W. L. Bragg (section 1.5.1), we denote the angle between $\mathbf{S}_0$ and $\mathbf{S}_1$ by 2θ and from simple geometry the magnitude of **s** is found as

$$s = 2\sin(\theta)/\lambda. \tag{1.6}$$

The scattering from P is shown in an Argand diagram in fig. 1.6; it is a complex quantity whose modulus is the amplitude of scattering, a, in the direction of $\mathbf{S}_1$ and whose argument, $2\pi \mathbf{r} \cdot \mathbf{s}$, represents the phase relative to that of a similar scatterer if placed at the origin, O. It should be noted that fig. 1.4 is three-dimensional and $\mathbf{r}, \mathbf{S}_0$ and $\mathbf{S}_1$ are not necessarily in one plane. However, of necessity $\mathbf{s}, \mathbf{S}_0$ and $\mathbf{S}_1$ are coplanar with **s** bisecting the angle between $\mathbf{S}_0$ and $\mathbf{S}_1$.

1.3.2 A distribution of point scatterers

The extension from a single point scatterer to a number of point scatterers is quite straightforward. If there are N scatterers the ith of which is at position

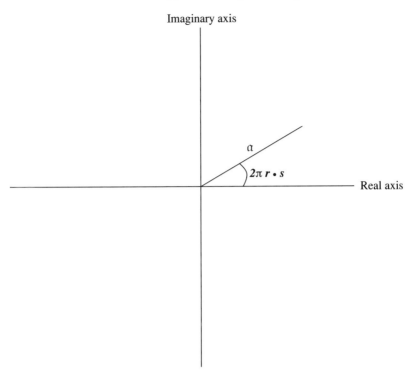

Fig. 1.6 Representation of the scattering from a single scatterer at vector position **r** as a complex quantity.

$\mathbf{r}_i$ with respect to the chosen origin and gives scattering amplitude a_i in the direction of $\mathbf{S}_1$ then the total scattering, as a complex quantity, will be

$$A(\mathbf{s}) = \sum_{i=1}^{N} a_i \exp(2\pi \mathbf{r}_i \cdot \mathbf{s}). \qquad (1.7)$$

This is shown in fig. 1.7 on an Argand diagram for $N = 4$. The intensity of the scattering in the direction $\mathbf{S}_1$ will be

$$I(\mathbf{s}) = |A(\mathbf{s})|^2, \qquad (1.8)$$

which is, of course, a real quantity.

1.3.3 Scattering from a continuous distribution

So far we have only considered scattering from point scatterers but there can also exist continuous distributions of *scattering material*. In fig. 1.8

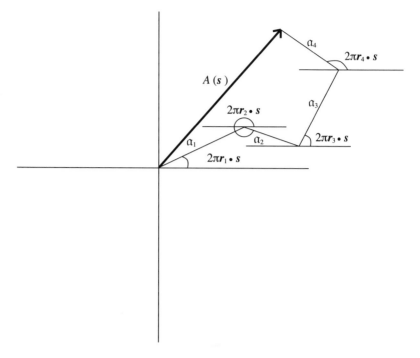

Fig. 1.7 The amplitude of scattering, $A(\mathbf{s})$, as the resultant of a distribution of four point scatterers at positions $\mathbf{r}_1$, $\mathbf{r}_2$, $\mathbf{r}_3$ and $\mathbf{r}_4$.

there is illustrated a general distribution of scattering material contained within some finite volume. The variation of scattering power from point to point may be expressed as $\sigma(\mathbf{r})$ which describes the scattered amplitude per unit volume in the direction $\mathbf{S}_1$. Thus with respect to the origin, O, the small element shown, of volume $\mathrm{d}V$, gives a contribution

$$\mathrm{d}A(\mathbf{s}) = \sigma(\mathbf{r}) \exp(2\pi\mathbf{r}\cdot\mathbf{s})\,\mathrm{d}V. \tag{1.9}$$

It is readily seen that the contribution of the total arrangement of scattering matter is given by

$$A(\mathbf{s}) = \int_V \sigma(\mathbf{r}) \exp(2\pi\mathbf{r}\cdot\mathbf{s})\,\mathrm{d}V \tag{1.10},$$

where the integral is over the total volume of the scattering material and the intensity of scattering is given by equation (1.8).

This concludes our generalised overview of the scattering process. Now we shall examine some specific applications pertaining to X-ray diffraction from crystals.

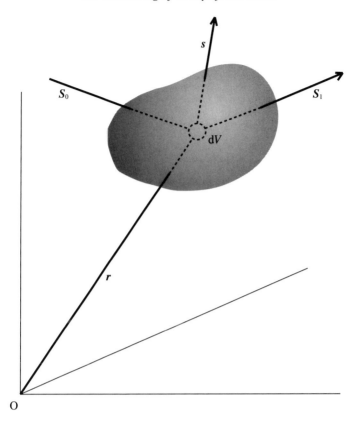

Fig. 1.8 Scattering from a small volume element of a distribution of scattering matter.

1.4 Scattering of X-rays from atoms

1.4.1 Scattering from a point electron

X-rays are a form of electromagnetic radiation and they are scattered through the interaction of their electric field component with atomic electrons. The form of the interaction of a high frequency electromagnetic wave with a free electron is easily understood in terms of a classical model. The electron is forced into oscillation with the same frequency as the incident electric field but π out of phase with it. Classical theory tells us that an oscillating, and hence accelerating, point charge will be a source of electromagnetic radiation, which travels outwards in all directions and constitutes the scattered wave. This kind of scattering is called Thomson scattering and the intensity of Thomson scattering from a free electron for an unpolarized incident beam of intensity I_0 is

$$I_{2\theta} = \tfrac{1}{2} I_0 \left(\frac{e^2}{4\pi\varepsilon_0 c^2 m} \right)^2 (1 + \cos^2 2\theta), \tag{1.11}$$

where e and m are the charge and mass of the electron, c is the speed of light and $I_{2\theta}$ is the energy of scattered radiation per unit solid angle at the scattering angle 2θ. The factor m^{-2} in the expression for $I_{2\theta}$ shows why scattering from nuclei is unimportant in this context.

1.4.2 Scattering from an isolated atom

We now consider X-ray scattering from the total electron content of an atom, which may be regarded as a spherically symmetric cloud of charge with the nucleus at its centre. The distribution is defined by $\rho(r)$, the electron density at distance r from the nucleus, and clearly the total electron content for an un-ionized atom

$$4\pi \int_0^\infty \rho(r) r^2 \, dr = z, \tag{1.12}$$

where z is the atomic number.

From equation (1.10) we can now write the total scattering from the electron cloud as

$$f(\mathbf{s}) = \int_{\substack{\text{all} \\ \text{space}}} \rho(r) \exp(2\pi \mathbf{r} \cdot \mathbf{s}) \, dV, \tag{1.13}$$

where $f(\mathbf{s})$ is the amplitude of scattering expressed in units of the scattering of a single point electron at the origin. The quantity $f(\mathbf{s})$ is called the *scattering factor* of the atom and because of the centrosymmetric electron-cloud distribution the scattering factor is real and the scattering is in phase with that from a point scatterer at the atomic centre.

The distributions $\rho(r)$ for various atoms can be found by quantum mechanical calculations, e.g. by Hartree, Hartree–Fock or Thomas–Fermi methods, and the resulting scattering factors are given for various values of s $(2\sin(\theta)/\lambda)$ in the *International Tables for Crystallography*. Some sample scattering factor curves are given in fig. 1.9; for a spherically symmetric distribution of electron density the scattering factor depends on the scalar

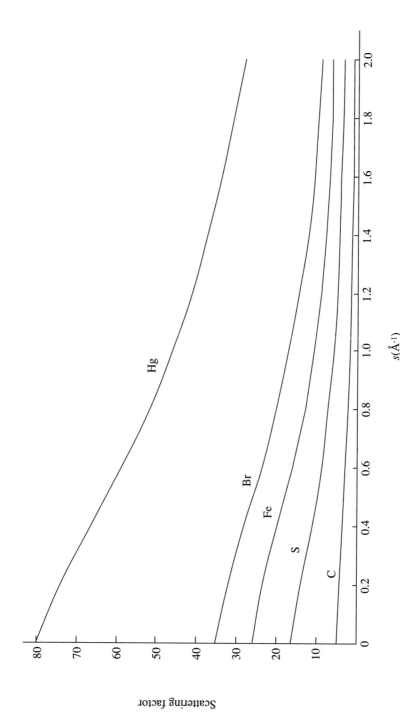

Fig. 1.9 Scattering factors as functions of s for a selection of atoms.

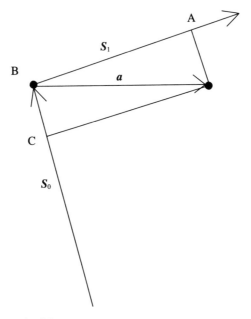

Fig. 1.10 The path difference from the incident to diffracted wavefront from neighbouring lattice points is $AB + BC = \mathbf{a} \cdot \mathbf{S} - \mathbf{a} \cdot \mathbf{S}_0$.

quantity, s, rather than the vector quantity, $\mathbf{s}$. From fig. 1.9 (and also from equation (1.13)) it will be seen that $f(0) = z$ and it will also be apparent that the f-curves for the heavier atoms fall off relatively less at higher scattering angles. The reason for the fall-off in scattering factor is the spread of the electron density which leads to interference between scattering from different parts of the cloud. Heavy atoms are more centrally condensed and approximate more closely to point atoms, which is why the fall-off is less severe.

1.5 X-ray diffraction from crystals

1.5.1 The reciprocal lattice

In fig. 1.10 are shown the incident and scattered beams from a point on a one-dimensional lattice of scatterers. Equation (1.3) gives the condition for constructive interference from neighbouring cells and from the figure it can be seen that, for cells separated in the **a** direction, this is equivalent to

$$\mathbf{a} \cdot (\mathbf{S}_1 - \mathbf{S}_0) = h\lambda$$

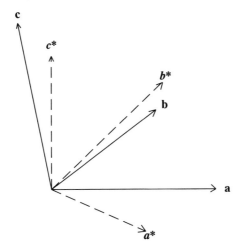

Fig. 1.11 Lattice vectors **a**, **b** and **c** and reciprocal-lattice vectors **a***, **b*** and **c***.

or, from equation (1.5)

$$\mathbf{a} \cdot \mathbf{s} = h.$$

Bringing in the other two dimensions the equations for a diffracted beam are

$$\mathbf{a} \cdot \mathbf{s} = h,$$
$$\mathbf{b} \cdot \mathbf{s} = k,$$
$$\mathbf{c} \cdot \mathbf{s} = l, \tag{1.14}$$

which are known as the Laue equations. We have seen that **s** is fixed by the incident and scattered beam directions but that, in general, for a particular direction of incident beam relative to the lattice no diffracted beam will occur.

The direction of an incident beam required to give a diffracted beam for indices (h, k, l) may be found from the concept of the reciprocal lattice. In fig. 1.11 are shown the lattice vectors **a**, **b** and **c** together with other vectors **a***, **b*** and **c*** that satisfy the conditions

$$\mathbf{a}^* \mathbf{a} = 1, \qquad \mathbf{a}^* \mathbf{b} = 0, \qquad \mathbf{a}^* \mathbf{c} = 0,$$
$$\mathbf{b}^* \mathbf{a} = 0, \qquad \mathbf{b}^* \mathbf{b} = 1, \qquad \mathbf{b}^* \mathbf{c} = 0,$$
$$\mathbf{c}^* \mathbf{a} = 0, \qquad \mathbf{c}^* \mathbf{b} = 0, \qquad \mathbf{c}^* \mathbf{c} = 1. \tag{1.15}$$

The last two equations in the first row give the direction of **a*** as perpendicular to the plane defined by **b** and **c**. The first equation then gives

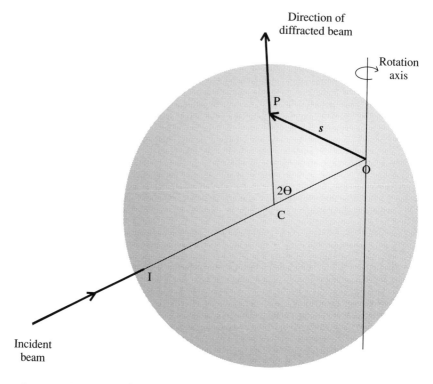

Fig. 1.12 The sphere of reflection. When the reciprocal-lattice point at position **s** touches the sphere the corresponding diffracted beam occurs.

the magnitude of **a***. The vectors **a***, **b*** and **c*** form the basis of the *reciprocal lattice*. It can be shown that the volume of a reciprocal lattice cell, V^*, equals V^{-1} where V is the volume of the cell defined by **a**, **b** and **c**.

We now consider the reciprocal lattice vector

$$\mathbf{s} = h\mathbf{a}^* + k\mathbf{b}^* + l\mathbf{c}^*. \qquad (1.16)$$

This vector satisfies the Laue equations (1.14) and so identifies the scattering vector **s**, which, in its turn gives the angle θ from equation (1.6). Any incident beam direction, $\mathbf{S}_0$, making an angle $(\pi/2 - \theta)$ with **s** (see fig. 1.5) will give rise to a diffracted beam in the plane defined by **s** and $\mathbf{S}_0$ and also making an angle $(\pi/2 - \theta)$ with **s**. The integers h, k, and l are called the *Miller indices* of the diffracted beam.

The condition under which diffraction will take place is illustrated in fig. 1.12, in which IO is the direction of a beam incident on a crystal at position O. With centre at C, on the line IO, a sphere is constructed of radius $1/\lambda$ such

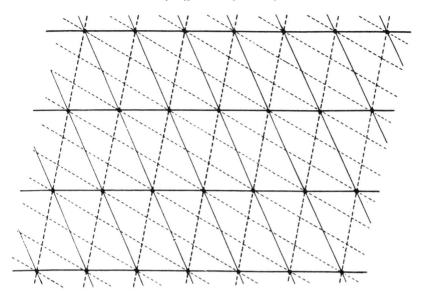

Fig. 1.13 A selection of sets of parallel equi-spaced lines on which lie all points of a two-dimensional array of scatterers.

that I and O are at opposite ends of a diameter. The end of the reciprocal lattice vector **s** touches the sphere at point **P**. It will be seen from the isosceles triangle COP and (1.6) that angle OCP equals 2θ and also that the incident beam makes an angle $\pi/2 - \theta$ with **s**. The condition for diffraction is satisfied with IC as the incident beam direction and CP as the direction of the diffracted beam. Thus, if a crystal is rotated about some axis, for example the one indicated in fig. 1.12, then whenever a reciprocal lattice point touches the sphere a diffracted beam will be produced.

The vectors $\mathbf{S_0}$, $\mathbf{S_1}$ and **s** can be interpreted in terms of specular reflection where $\mathbf{S_0}$ and $\mathbf{S_1}$ are the incident and reflected beams respectively and **s** is the normal to the reflecting plane. An early interpretation of X-ray diffraction by W. L. Bragg was as reflection from richly populated layers of atoms in a crystal. In fig. 1.13 there is shown a simple two-dimensional array of scatterers, all of which fall on sets of equi-spaced straight lines; in three dimensions all scatterers forming a simple three-dimensional array would fall on sets of equi-spaced planes. According to the Bragg model each of these planes would act as a partial reflector of X-rays. When parallel rays in an incident beam, corresponding to a plane wavefront, fall on a specularly reflecting surface then after reflection from any two points on the surface they will still be in phase. This is because AD = CB in fig. 1.14(a). However,

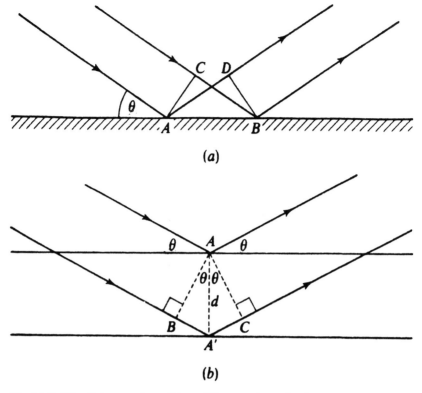

Fig. 1.14(a) Parallel rays reflected from different points of a plane are in phase after reflection. (b) Parallel rays reflected from points on neighbouring partially reflecting planes are in phase when Bragg's law is obeyed.

when there are partially-reflecting planes then there must be the further condition that reflection from neighbouring planes must also be in phase. From fig. 1.14(b) it is clear that the condition that reflection from points A and A' are in phase is that

$$BA' + A'C = 2d\sin\theta = \lambda. \qquad (1.17)$$

This expression is known as Bragg's law and it gives the angle of reflection, θ, in terms of the lattice spacing, d, and the X-ray wavelength, λ. It will be seen that λ on the right-hand side of (1.17) could be replaced by $n\lambda$ but this would be equivalent to having a spacing d/n, which is also a set of planes on which all the scatterers fall, albeit that some planes contain no scatterers. If $n = 1$ corresponds to Miller indices (h, k, l) then $n = 2$ will correspond to the reflection with Miller indices $(2h, 2k, 2l)$.

The Bragg description of diffraction has very much influenced the terminology of X-ray crystallography. Thus diffracted beams are referred to as *reflections* and the sphere shown in fig. 1.12 is known as the *sphere of reflection*.

1.5.2 The structure-factor equation

The model we have of a crystal is that of a three-dimensional array of parallelepiped-shaped unit cells each of which contains a similar arrangement of atoms. Such a body will diffract X-rays according to the conditions of the Laue equations and the amplitude of the diffracted beam with indices (h,k,l) will be proportional to the amplitude of scattering in the same direction by the contents of one unit cell. We can consider the contents of a unit cell in two ways. In the first way we regard the atoms as providing a general distribution of scattering (electron) density and we use equation (1.10) to give

$$F(h,k,l) = \int_{\text{cell}} \rho(r) \exp(2\pi \mathbf{r} \cdot \mathbf{s}) \, dV, \tag{1.18}$$

where $F(h,k,l)$ is called the *structure factor* for the associated reflection.

Another way of looking at the cell's contents is to think of the individual atoms situated at points in the unit cell, each atom having a scattering power given by the appropriate scattering factor. It is customary to use fractional coordinates (x,y,z) so that a position in the unit cell can be written as

$$\mathbf{r} = x\mathbf{a} + y\mathbf{b} + z\mathbf{c}, \tag{1.19}$$

where x, y and z are all in the range 0–1. From the definition for s, equation (1.16), and the definitions of reciprocal lattice vectors, equation (1.15), it is found that

$$\mathbf{r} \cdot \mathbf{s} = hx + ky + lz. \tag{1.20}$$

If the unit cell contains N atoms, the jth one of which has fractional coordinates (x_j, y_j, z_j), then from equation (1.7)

$$F(h,k,l) = \sum_{j=1}^{N} f_j \exp\left[2\pi i (hx_j + ky_j + lz_j)\right]. \tag{1.21}$$

In general the structure factor is a complex quantity so that we may write

$$F(h,k,l) = |F(h,k,l)| \exp\left[i\phi(h,k,l)\right], \tag{1.22}$$

where $|F(h,k,l)|$ is the *structure amplitude* and $\phi(h,k,l)$ is the *phase* of the reflection. Many crystals have a centre of symmetry, which would conventionally be taken as the origin for the atomic coordinates. Then to every (x_j, y_j, z_j) in equation (1.21) there is also a $(-x_j, -y_j, -z_j)$ with the same f_j so that for each term there is a complex conjugate term and the structure factor is real, corresponding to a phase of 0 or π. The phase factor, $\exp(i\phi)$, is thus either $+1$ or -1 and for this reason it is usual to refer to the *sign* rather than the phase of a centrosymmetric structure factor. For a centrosymmetric structure the structure-factor equation becomes

$$F(h,k,l) = \sum_{j=1}^{N} f_j \cos\left[2\pi(hx_j + ky_j + lz_j)\right]. \tag{1.23}$$

Another point worth noting is the relationship between the structure factor of index (h,k,l) and that of index $(-h,-k,-l)$, or $(\bar{h},\bar{k},\bar{l})$ in the more usual crystallographic notation. It will be seen from equation (1.21) that $F(\bar{h},\bar{k},\bar{l})$ is the complex conjugate of $F(h,k,l)$ so that

$$|F(\bar{h},\bar{k},\bar{l})| = |F(h,k,l)|,$$
$$\phi(\bar{h},\bar{k},\bar{l})| = -\phi(h,k,l). \tag{1.24}$$

The intensity of the (h,k,l) reflection will be given by

$$I(h,k,l) = C \times |F(h,k,l)|^2, \tag{1.25}$$

where C depends upon various physical factors that influence the intensities measured by X-ray experiments. One relationship between intensities that we can see from (1.24) and (1.25) is that

$$I(h,k,l) = I(\bar{h},\bar{k},\bar{l}), \tag{1.26}$$

which is known as Friedel's law. In some circumstances Friedel's law can break down and one of the principal topics of this book will be the exploitation of this breakdown as an aid to solving crystal structures.

1.5.3 Factors affecting observed intensities

A full treatment of the various factors influencing the observed intensities of X-ray reflections is beyond the scope of this book. Fairly detailed mathematical and physical descriptions will be found in various specialized textbooks (e.g. Woolfson, 1970). What will be done here is to give the physical basis of each factor and its mathematical form where appropriate.

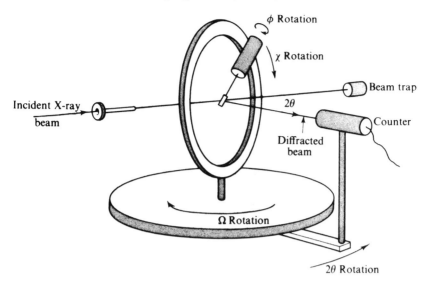

Fig. 1.15 A typical four-circle diffractometer. The counter rotates about the 2θ axis in one plane and the crystal may be oriented in any direction by the three axes of rotation ϕ, χ and Ω.

1.5.3.1 The geometry of the diffracting process

It is normal in a diffraction experiment to collect data from a rotating crystal. For example, a computer-controlled four-circle diffractometer, as shown in fig. 1.15, may be used. The Ω, χ and ϕ circles are adjusted to place the crystal in a position to diffract the beam with indices (h,k,l) in the horizontal plane and the 2θ circle places the photon counter in a position to receive the diffracted beam. A common strategy is to collect the diffracted photons in a $\Omega - 2\theta$ scan, which means that the crystal and counter are both rotated about the vertical axis with the 2θ circle moving at twice the speed of the Ω circle.

When data are being collected photographically from a rotating crystal the total energy in the diffracted beam is given by

$$E(h,k,l) = \frac{\lambda^3}{\omega} I_0 \left(\frac{e^2}{4\pi\varepsilon_0 c^2 m} \right)^2 \frac{1+\cos^2(2\theta)}{2\sin(2\theta)} \frac{V_x}{V^2} |F(h,k,l)|^2, \quad (1.27)$$

where λ is the wavelength of the X-rays, ω the angular speed of rotation of the crystal, V_x the volume of the crystal and V the volume of the unit cell. The terms involving Thomson scattering and the structure factor can be

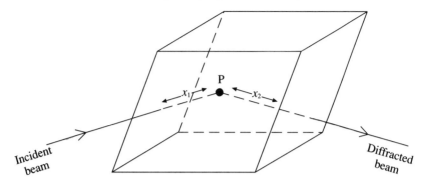

Fig. 1.16 The X-ray beam scattered by the material at point P has a total path $x_1 + x_2$
in the crystal.

seen to be present, as is the volume of the crystal, which is a measure of the
total amount of diffracting material available. The speed of rotation
directly affects the length of time the crystal spends in the diffracting
position so that, for example, if ω is halved then the crystal will diffract for
twice as long. The term $\frac{1}{2}[1 + \cos^2(2\theta)]$ is called the *polarization factor* and
assumes that the X-ray beam is unpolarized. The other trigonometric term
$1/\sin(2\theta)$ is called the *Lorentz factor* and depends on the diffraction
geometry. The last two terms, lumped together and referred to as the *Lp
factor*, may be found in tables. The remaining factors, involving λ and V,
deal with the effect of the spread of the diffracting region around the precise
reciprocal lattice position given by the Laue equations.

1.5.3.2 The absorption of X-rays

When X-rays pass through matter the intensity is attenuated according to
the law

$$I = I_0 \exp(-\mu x), \tag{1.28}$$

where I_0 is the incident intensity and I is the intensity after passing through a
distance x of material with a *linear absorption coefficient* μ. In fig. 1.16 we
show the path of an incident ray diffracted from a small region of a crystal.
The intensity of the emerging beam will be lowered by a factor of
$\exp[-\mu(x_1 + x_2)]$ due to absorption. The lowering of the intensity of
scattering from the total crystal is found by aggregating the absorption
effect from all regions of the crystal.

At the microscopic level absorption occurs through the interaction of X-
rays with individual atoms either through giving them vibrational motion,

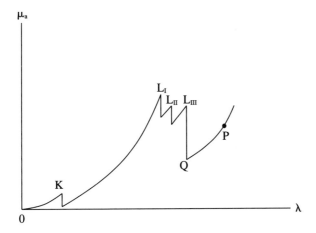

Fig. 1.17 Variation of atomic absorption coefficient with wavelength for a typical element.

which appears as temperature, or by the Compton effect in which the photon has an elastic collision with an atomic electron. If X-rays passed through a material consisting of a single atomic species then the attenuation of the beam would be given by

$$\frac{\mathrm{d}I}{I} = -\mu_a \,\mathrm{d}n, \tag{1.29}$$

where $\mathrm{d}n$ is the number of atoms in the path of unit area of the beam and μ_a is the *atomic absorption coefficient* for the element in question. The way in which μ_a varies with the wavelength of the X-rays is shown in fig. 1.17. Starting from the point P the absorption reduces as the wavelength is reduced, corresponding to the radiation becoming more penetrating. However, at the point Q the absorption abruptly increases. The reason for this is that a new mechanism for losing energy has become available since, at this wavelength, a photon has enough energy, hc/λ, to eject an atomic electron – in this case one in the L_{III} shell. These discontinuities in the atomic absorption coefficient are known as *absorption edges* and they have relevance to the phenomenon of anomalous scattering (chapters 4 and 6), which underlies an important technique for solving crystal structures.

1.5.3.3 Extinction effects

There are two main types of extinction effect that reduce the intensity of an X-ray reflection below that which it would be otherwise. The first, *primary extinction*, is due to dynamical effects where there is interference between beams of radiation that have undergone multiple scattering within the crystal. This effect is significant for very intense reflections but it can be reduced by having a mosaic crystal consisting of a collection of small slightly misaligned blocks for which there is no coherence in scattering from different blocks. The physical processes giving rise to primary extinction will be dealt with in more detail in §8.2.

The other type of extinction, *secondary extinction*, amounts to an increase in the effective absorption coefficient due to energy being extracted from the incident beam by diffraction. Absorption effects can be reduced by the use of small crystals and this is also an effective strategy for reducing secondary extinction. However, the penalty for this is a weaker diffracted beam, which entails either a loss of accuracy or an increase in the time of data collection.

1.5.3.4 *The temperature factor*

We have seen that the scattering factor of an atom is less than that of a superposition of point electrons because the spread of the electron density leads to interference between scattering from different parts of the cloud. There is an additional contribution to the effective spread of the electron density – the temperature of the crystal, which manifests itself on the atomic scale as the vibration of atoms. The time-averaged view of a vibrating atom gives a more diffuse distribution of electron density than does a stationary one, for which the tabulated scattering factors apply.

The simplest case to consider is where the vibration is isotropic so that the time-averaged electron density is spherically symmetric. In this case the effective scattering factor becomes

$$f_T = f \exp\left[-B\sin^2(\theta)/\lambda^2\right] \qquad (1.30)$$

where B is the isotropic *temperature factor*. Sometimes it is a sufficiently good approximation to assume not only that the thermal vibrations are isotropic for all atoms but also that all atoms have the same value of B. This would give the temperature-modified structure factor as

$$F(h,k,l)_T = F(h,k,l)\exp\left[-B\sin^2(\theta)/\lambda^2\right]. \qquad (1.31)$$

The factor by which the intensity is reduced by thermal vibration, which will be $\exp\left[-2B\sin^2(\theta)/\lambda^2\right]$ in this case, is known as the *Debye–Waller factor*.

For the most refined work it is necessary to take into account not only that atoms have different temperature factors but also that they may not be isotropic. For an anisotropic thermal vibration, contours of constant time-averaged electron density are ellipsoids and the temperature factor for a given atom and given reflection will be

$$B_j(h,k,l) = 8\pi \overline{u(\mathbf{s})}_j^2, \qquad (1.32)$$

where $\overline{u(\mathbf{s})}_j^2$ is the mean square displacement of atom j along the direction of $\mathbf{s}$.

This concludes our review of the factors influencing the observed intensities of X-ray reflections. There is, in fact, one more important factor, anomalous scattering, to which reference has previously been made with respect to absorption edges.

1.5.4 The electron density equation

We have seen from equation (1.21) that a knowledge of the crystal structure, i.e. the positions of all the atoms, enables structure factors to be calculated. Now we consider the inverse problem, that is, how do we determine the electron density given the structure factors?

The most convenient approach to this problem is via Fourier transform theory. This tells us that if we have two functions $f(\mathbf{r})$ and $F(\mathbf{s})$ defined in two reciprocally related spaces such that

$$F(\mathbf{s}) = \int_V f(\mathbf{r}) \exp(2\pi i \mathbf{r}\cdot\mathbf{s}) \, dV \qquad (1.33)$$

then

$$f(\mathbf{r}) = \int_{V^*} F(\mathbf{s}) \exp(-2\pi i \mathbf{r}\cdot\mathbf{s}) \, dV^* \qquad (1.34)$$

where the integrals are over the whole of the volumes V and V^* for which $f(\mathbf{r})$ and $F(\mathbf{s})$ are non-zero respectively. Comparing equations (1.33) and (1.18) we identify $F(\mathbf{s})$ with $F(h,k,l)$ and $f(\mathbf{r})$ with $\rho(\mathbf{r})$. We also note that $F(h,k,l)$ is a discrete function at reciprocal lattice points and each F is associated with a volume $V^* = V^{-1}$. This leads to the electron-density equation

$$\rho(\mathbf{r}) = \frac{1}{V} \sum_h \sum_k \sum_l F(h,k,l) \exp(-2\pi i \mathbf{r}\cdot\mathbf{s}) \qquad (1.35)$$

or

$$\rho(x,y,z) = \frac{1}{V} \sum_h \sum_k \sum_l F(h,k,l) \exp\left[-2\pi i(hx+ky+lz)\right] \qquad (1.36)$$

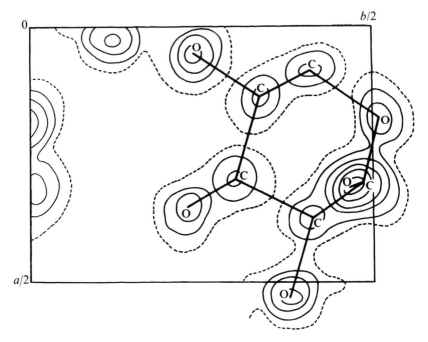

Fig. 1.18 The electron density of D-xylose projected down the *c*-axis. The dashed contour is at 3 e Å^{-2} and others at intervals of 2 e Å^{-2} (Woolfson, 1958).

Within the summation of (1.36) there will be pairs of terms with indices (h,k,l) and $(\bar{h},\bar{k},\bar{l})$, which will contribute

$$|F(h,k,l)|\exp\left[i\phi(h,k,l)\right]\exp\left[-2\pi i(hx+ky+lz)\right]$$
$$+|F(\bar{h},\bar{k},\bar{l})|\exp\left[i\phi(\bar{h},\bar{k},\bar{l})\right]\exp\left[2\pi i(hx+ky+lz)\right].$$

From the relationships (1.24) it can be found that the contribution of the pair of terms is

$$2|F(h,k,l)|\cos\left[2\pi(hx+ky+lz)-\phi(h,k,l)\right].$$

Using this result the electron-density equation (1.36) becomes

$$\rho(x,y,z)=\frac{1}{V}\sum_{h}\sum_{k}\sum_{l}|F(h,k,l)|\cos\left[2\pi(hx+ky+lz)-\phi(h,k,l)\right]. \qquad (1.37)$$

Electron density maps are usually computed at points on a three-dimensional grid. From the interpolated peaks of the density it is possible to deduce the positions of the atomic centres and hence to solve the crystal structure. For illustration, fig. 1.18 shows a two-dimensional electron density map projected along the *c* axis for the structure of the sugar, D-xylose. Contours of constant electron density clearly show the positions of the individual atoms. However, finding the electron density is not a straightforward matter. While the values of the $|F|$ terms can be found from the observed intensities, the values of the phases cannot immediately be found from the data. A review of the various ways of solving crystal structures, which are subsequently dealt with in this book, is given in §1.7.

1.6 The symmetry of crystals

1.6.1 The crystal lattice

A crystal is a three-dimensional periodic object and we have seen in §1.2 how this gives rise to discrete diffracted beams. One way to see the periodicity is by taking a point in one of the unit cells together with the equivalent points in all other cells. Such a set of points will form a lattice and there are seven *primitive lattices* based on the seven possible *crystal systems*. A general unit cell is shown in fig. 1.19(a); it is a parallelepiped based on the lengths of three cell edges and three angles. In table 1.1 there are given the characteristics of the unit cell in the seven crystal systems. A representation of a primitive orthorhombic lattice is given in fig. 1.19(b) although it must be realized that the actual lattice is infinite in extent.

There is a type of symmetry that a crystal may possess, which is intimately linked with the lattice itself. This is illustrated in fig. 1.19(c) with the lattices orthorhombic C, orthorhombic I and orthorhombic F. In the orthorhombic C lattice (with C-face centring) for each point in the cell there is an exactly equivalent point with a displacement $\frac{1}{2}\mathbf{a} + \frac{1}{2}\mathbf{b}$. It would be possible to construct a primitive lattice from the equivalent points but this would conceal the orthorhombic nature of the arrangement so this is not done. Orthorhombic I corresponds to body centring where for each point there is an equivalent point with displacement $\frac{1}{2}(\mathbf{a} + \mathbf{b} + \mathbf{c})$ and orthorhombic F corresponds to simultaneous A, B and C-face centring. The right-hand column of table 1.1 lists the fourteen possible *Bravais lattices*. No others can occur.

1.6.2 Symmetry elements

The atomic arrangement in a crystal can possess symmetries of various kinds, which are described in terms of *symmetry elements*. The possible symmetry elements of a crystal structure will now be given, with figures where appropriate.

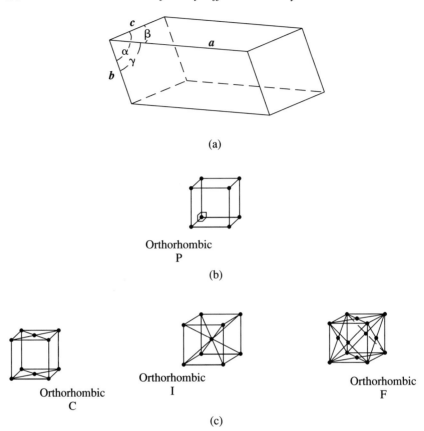

(a)

Orthorhombic
P

(b)

Orthorhombic
C

Orthorhombic
I

Orthorhombic
F

(c)

Fig. 1.19(a) A general unit cell based on edges **a**, **b** and **c**. (b) Representation of a primitive orthorhombic lattice. (c) Representation of non-primitive orthorhombic lattices C, I and F.

1.6.2.1 Centre of symmetry ($\bar{1}$)

This is a point such that if there is an atom in the cell with vector displacement **r** from the centre of symmetry then there is another atom with dispacement $-$**r**. The unit cell in fig. 1.20(a) has a centre of symmetry at the origin. Other centres of symmetry automatically occur although it should be noted that they are not equivalent to each other or to the one at the origin. If there is a group of objects related to another group by a centre of symmetry then the groups will be enantiomorphs, i.e. related in symmetry (but not necessarily spatially) as an object and its mirror image. A left hand and a right hand are so related; they can so be held as to be related by a

Table 1.1. *The seven crystal systems and fourteen Bravais lattices*

System	Unit cell	Lattices
Triclinic	No special relationship	P
Monoclinic	$a \neq b \neq c$	
	$\beta \neq \alpha = \gamma = 90°$	P C
Orthorhombic	$a \neq b \neq c$	
	$\alpha = \beta = \gamma = 90°$	P C I F
Tetragonal	$a = b \neq c$	
	$\alpha = \beta = \gamma = 90°$	P I
Trigonal	$a = b = c$	
	$\alpha = \beta = \gamma \neq 90°$	R ($\equiv$P)
Hexagonal	$a = b \neq c$	
	$\alpha = \beta = 90°; \gamma = 120°$	P
Cubic	$a = b = c$	
	$\alpha = \beta = \gamma = 90°$	P I F

mirror plane but equally they can be placed so as to be related by a centre of symmetry.

1.6.2.2 Mirror plane (m)

The mirror planes across two equivalent opposite faces in fig. 1.20(b) together with the periodicity of the cell generate from atom A_1 other atoms at A_2, A_1' and A_2'. A_1 and A_2' are then related by another mirror plane, which bisects the original pair as shown.

1.6.2.3 Glide planes (a, b, c, n, d)

These symmetry elements operate by a combination of reflection and translation. For an *a*-glide plane perpendicular to the **b** direction a new point is generated by first reflecting across the plane and then translating by $\frac{1}{2}$**a**. For an *n*-glide plane perpendicular to **a**, generation of a new point consists of reflecting across the plane and then displacing by $\frac{1}{2}$(**b** + **c**). As is the case for the mirror plane, two sets of interleaved glide planes are generated.

The *d*-glide plane is complicated and quite rare and will not be dealt with here.

1.6.2.4 Rotation axes (2,3,4,6)

An *n*-fold rotation axis is one for which new points are generated by a rotation of $2\pi/n$ about the axis. A fourfold axis, based on a tetragonal unit

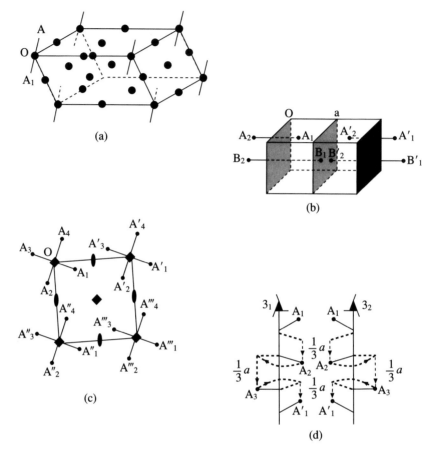

Fig. 1.20(a) A centre of symmetry, $\bar{1}$. (b) A mirror plane, m. (c) A fourfold axis, 4. (d) The operation of 3_1 and 3_2 screw axes.

cell, is illustrated in fig. 1.20(c); it will be seen that a system of twofold axes is also generated.

1.6.2.5 Screw axes $(2_1; 3_1, 3_2; 4_1, 4_2, 4_3; 6_1, 6_2, 6_3, 6_4, 6_5)$

The action of a R_m screw axis along the **b** direction is a rotation $2\pi/R$ followed by a translation $(m/R)\mathbf{b}$. The operations of 3_1 and 3_2 axes are shown in fig. 1.20(d). The set of points produced are enantiomorphically related.

1.6.2.6 Inversion axes $(\bar{3}, \bar{4}, \bar{6})$

The action of the inversion axis $\bar{R}$ is to rotate the point about the axis by an angle $2\pi/R$ followed by inversion through a point contained in the axis.

Table 1.2. *Symmetry elements and the symbols that represent them*

Type of symmetry element	Written symbol	Graphical symbol	
Centre of symmetry	$\bar{1}$	o	
		Perpendicular to paper	In plane of paper
Mirror plane	*m*	————	⌐ ⟍
Glide planes	*a b c*	‑ ‑ ‑ ‑ ‑ ‑	← ⌐↓
		Glide in plane of paper	Arrow shows glide direction
		· · · · · · · · · · Glide out of plane of paper	
	n	‑·‑·‑·‑·‑	↗
Rotation	2	⬤	———→
	3	▲	
	4	◆	
	6	⬢	
Screw axes	2_1	⬤	———→
	$3_1, 3_2$	▲ ▲	
	$4_1, 4_2, 4_3$	◆ ◆ ◆	
	$6_1, 6_2, 6_3, 6_4, 6_5$	⬢ ⬢ ⬢ ⬢ ⬢	
Inversion axes	$\bar{3}$	▲	
	$\bar{4}$	◈	
	$\bar{6}$	⬡	

A list of the symmetry elements that have been described and the graphical symbols used to represent them are given in table 1.2.

1.6.3 Space groups

For any particular crystal there may be combinations of the symmetry elements that have been described. There are 230 distinctive arrangements of symmetry elements which might occur, each of which is called a *space*

Triclinic 1̄ $P\bar{1}$

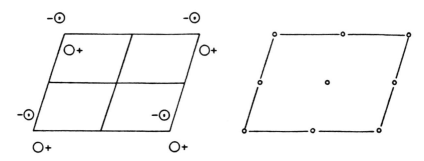

Fig. 1.21 Representation of the space group P1̄.

group. There will now be described four common space groups, which will illustrate the nomenclature and the way in which they are represented by diagrams.

1.6.3.1 Triclinic P1̄

This space group is based on a primitive triclinic cell that contains a centre of symmetry. It is represented in fig. 1.21 in the form given in the *International Tables for X-ray Crystallography*. The circle represents a structural unit; the symbols + and − beside them indicate that the coordinates out of the paper are of the form $+t$ and $-t$. A contained comma indicates that the structural unit is enantiomorphically related to the unit without a comma. The right-hand diagram shows the set of symmetry elements within the cell.

1.6.3.2 Monoclinic P2₁

This space group, illustrated in fig. 1.22, involves a primitive monoclinic cell with twofold screw axes along **c**, the cell edge that is perpendicular to the other two. In this case $\frac{1}{2}+$ associated with a circle means that the coordinate is $\frac{1}{2}+z$. It may be noticed that, while there are no centres of symmetry, the projection down **c** of a structure with this space group will be centrosymmetric.

1.6.3.3 Monoclinic C2

This space group, projected down the **c** direction, is illustrated in fig. 1.23. The plan view shows the C-face centring quite clearly. For each structural unit with a particular z value there is a similar unit with the same z coordinate displaced by $\frac{1}{2}(\mathbf{a}+\mathbf{b})$. The twofold axes, which are at the level

Monoclinic $P2_1$

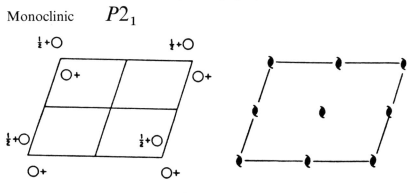

Origin on 2_1; unique axis c

Fig. 1.22 Representation of the space group $P2_1$.

Monoclinic $C2$

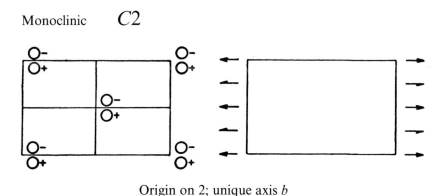

Origin on 2; unique axis b

Fig. 1.23 Representation of the space group C2.

$z = 0$, are shown by the full-headed arrows. They relate the close pairs of units with coordinates $+z$ and $-z$. Although it is not part of the description of the space group, twofold screw axes, shown by the half-headed arrows, are also generated. A careful examination of the diagram will show the structural units related by the 2_1 axes.

1.6.3.4 Orthorhombic $P2_12_12_1$

This space group is a very commonly occurring one with natural products and is well worth careful study. It is based on a primitive orthorhombic cell and it has three sets of orthogonal non-intersecting 2_1 axes. If either two or all three of the sets of screw axes intersect then other symmetry elements are

Orthorhombic $P2_12_12_1$

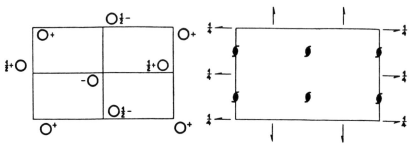

Origin halfway between three pairs of non-intersecting screw axes

Fig. 1.24 Representation of the space group $P2_12_12_1$.

produced and the space group is something different. In fig. 1.24 the arrangements of structural units and of symmetry elements are shown. The label $\frac{1}{4}$ attached to the horizontal axes in the plane of the diagram gives their height above the plane of the paper. In fact other ones also appear at height $\frac{3}{4}$.

This concludes our description of symmetry elements and space groups. For a full description of all 230 space groups the reader is referred to *International Tables for X-ray Crystallography*, Volume A, published by D. Reidel on behalf of the International Union of Crystallography.

1.7 Methods of solving crystal structures

The basic problem of structural crystallography is that of determining crystal structures from the observed intensities. Equation (1.37) encapsulates the essence of this problem since experiment gives everything needed to solve the structure except phases; this is referred to as the *phase problem in crystallography*. In this book we shall be covering all the common methods but they can be categorised into two types.

In the first class of method data are collected from a crystal of interest by a straightforward diffraction experiment and this single data set is used to solve the structure. The most basic method, used for the very earliest solutions, was the *trial-and-error method*, which involved trying various proposed structures until one was found that was consistent with the observed intensities. The great breakthrough that began the process of making X-ray crystallography a major scientific tool was the development of *Patterson-map methods* (chapter 2), which depend only on the measured intensities and give information about interatomic vectors in the structure. The final, and the most important, of this first class of methods are the so-called *direct methods* (chapter 3), which directly determine phases by mathematical procedures from the set of observed intensities.

In the second class of method, which we categorise as *physical methods*, there is some departure from a single simple diffraction experiment. In the *isomorphous-replacement method* there are used the data sets of two or more different, but related, crystals, which have the same crystal structure except that some atoms are of different kinds or where some atoms present in one crystal are completely absent in another. Another technique, of increasing importance, is the *anomalous scattering method* where the wavelength of the diffracted radiation is adjusted to give a breakdown of Friedel's law (anomalous scattering) with phase information being derived from the different values of the Friedel pairs $|F(h, k, l)|$ and $|F(\bar{h}, \bar{k}, \bar{l})|$. An introduction to both isomorphous replacement and anomalous scattering methods is given in chapter 4 and more advanced aspects of these methods are developed in chapters 5 and 6.

A class of physical method, completely different from any other so far mentioned, involves a combination of high-resolution electron microscopy with electron diffraction. This is important for the structure analysis of microcrystalline materials and is described in chapter 7.

The most advanced applications of modern methods of solving crystal structures are made with protein structures and with methods based on both isomorphous replacement and anomalous scattering the outcome is often a set of phase estimates but only for reflections of limited resolution – i.e. up to a maximum value of $(\sin \theta)/\lambda$. To complete the structure determination then requires a process of *phase extension and refinement* and chapter 7 deals with this topic.

Another type of physical method, still in an active state of development, are the *multi-beam reflection methods*. These are based on deriving phase information from experimental procedures and are described in chapter 8.

2

The Patterson and heavy-atom methods

2.1 The Patterson function

2.1.1 The basic theory

Early attempts to solve crystal structures were usually based on a trial-and-error process. With a very simple structure and a unit cell containing symmetry elements, on or about which atoms had to be situated, it is sometimes possible to restrict the number of possible atomic arrangements. Each possible arrangement is then a trial structure for which calculated structure factors, $F_c(h,k,l)$, may be found and whose magnitudes may then be compared with the observed structure amplitudes, $|F_o(h,k,l)|$. The measure of agreement is the *residual* or *reliability index*

$$R = \frac{\sum_{\mathbf{h}} \left| |F_o(\mathbf{h})| - |F_c(\mathbf{h})| \right|}{\sum_{\mathbf{h}} |F_o(\mathbf{h})|}, \tag{2.1}$$

where $\mathbf{h}$ represents the vector index (h,k,l). A low value of R would be taken as an indication that the trial structure was approximately correct and then a process of structure refinement could begin. An electron density map would be calculated from equation (1.36) in which were used the observed structure amplitudes, $|F_o(\mathbf{h})|$, and the calculated phases, $\phi_c(\mathbf{h})$. The peaks of this map would give revised, and presumably better, trial atomic positions from which new calculated structure factors could be found. The steps of electron-density and structure-factor calculation would then be repeated until the process converged to stable atomic positions.

The process of solving structures was revolutionised when Patterson (1934) proposed the use of the function

$$P(\mathbf{r}) = \frac{1}{V} \sum |F(\mathbf{h})|^2 \cos(2\pi \mathbf{h} \cdot \mathbf{r}). \tag{2.2}$$

We should note that $\mathbf{h}$ represents a position in reciprocal space equivalent to that of $\mathbf{s}$ in equation (1.16) so that, as indicated in equation (1.20),

$$\mathbf{h} \cdot \mathbf{r} = hx + ky + lz. \tag{2.3}$$

The *Patterson function* gives a map, which shows peaks at the positions of interatomic vectors with each peak having a weight proportional to the product of the atomic numbers of the two atoms. Starting with the structure-factor equation written in the form

$$F(\mathbf{h}) = \sum_{j=1}^{N} f_j \exp(2\pi i \mathbf{h} \cdot \mathbf{r}_j) \tag{2.4}$$

then, since $F(\bar{\mathbf{h}})$ is the complex conjugate of $F(\mathbf{h})$,

$$F(\bar{\mathbf{h}}) = \sum_{k=1}^{N} f_k \exp(-2\pi i \mathbf{h} \cdot \mathbf{r}_k). \tag{2.5}$$

Combining equations (2.4) and (2.5)

$$F(\mathbf{h})F(\bar{\mathbf{h}}) = |F(\mathbf{h})|^2 = \sum_{j=1}^{N} \sum_{k=1}^{N} f_j f_k \exp[2\pi i \mathbf{h} \cdot (\mathbf{r}_j - \mathbf{r}_k)]. \tag{2.6}$$

Comparing equations (2.6) and (1.21) it can be seen that $|F(\mathbf{h})|^2$ plays the role of a structure factor for a structure with scatterers at positions $\mathbf{r}_j - \mathbf{r}_k$ with scattering factors $f_j f_k$. The corresponding density equation to (1.35) then becomes

$$P(\mathbf{r}) = \frac{1}{V} \sum_{\mathbf{h}} |F(\mathbf{h})|^2 \exp(-2\pi i \mathbf{h} \cdot \mathbf{r}) \tag{2.7}$$

but, because of the existence of pairs of terms in the summation with indices $\mathbf{h}$ and $\bar{\mathbf{h}}$, only the real part of the summation remains to give equation (2.2).

Let us consider what will be seen in a Patterson map for a structure containing N atoms in positions $\mathbf{r}_i (i = 1 - N)$. There will be the N null vectors between each atom and itself, which contribute to a large peak at the origin. Then there will be $N(N-1)$ general peaks $(\mathbf{r}_j - \mathbf{r}_k)$, which will occur in centrosymmetric pairs since for each (j, k) there is a (k, j). In fig. 2.1 there is shown a simple structure and a representation of the vector set. Since there are approximately N^2 Patterson peaks in the main body of the unit cell it is obvious that there is a great deal of overlap and in general, even for

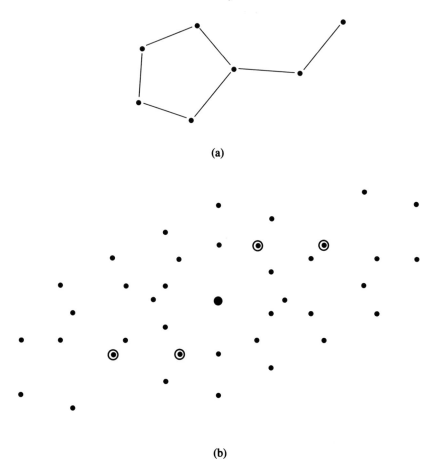

Fig. 2.1(a) The structure. (b) The vector set. Points marked ⊙ are double vectors and ● is the origin point.

structures of moderate size, it is difficult to pick out many individual vectors.

2.1.2 Parallel vectors and heavy atoms

There are circumstances in which interatomic vectors can be identified. In fig. 2.2 there is shown the structure of anthracene, a planar molecule, and it will be seen that there will be many overlapping parallel vectors, for example those shown by heavy lines. Where there is a systematic overlap of vectors then it may be possible to pick out this feature against the

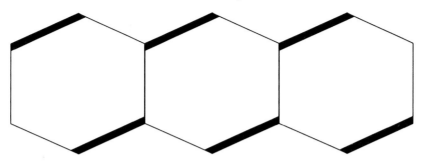

Fig. 2.2 The carbon atoms of the anthracene molecule. Heavy lines indicate a repeated vector.

background of peaks of lesser height. As an example of this, in fig. 2.3(a) there is shown the arrangement of molecules in the structure of 4,6-dimethyl-2-hydroxypyrimidine (Pitt, 1948) projected down the *a* axis. All the rings are approximately parallel and the vectors AC, BD and CE occur repeatedly. The two-dimensional Patterson projection, fig. 2.3(b), computed with coefficients $|F(0,k,l)|^2$, shows the position of these vectors at positions X, Y and Z respectively. From the Patterson projection the approximate orientation of the rings may be determined without ambiguity although the position of the hydroxy group needs to be fixed for a complete determination of the molecular orientation.

Another case in which interatomic vectors can be located is when a structure contains a small number of heavy atoms. The total weight contained in the Patterson peak linking two atoms is proportional to the product of the atomic numbers. If, for example, a structure contains carbon and chlorine atoms, of atomic numbers six and seventeen respectively, then the relative weights of C–C, C–Cl and Cl–Cl peaks are 36:102:289. As long as the number of carbon atoms is not too large it should be possible to pick out the peaks linking chlorine atoms and to deduce the chlorine substructure. This is illustrated with the structure of 2-amino-4,6-dichloropyrimidine (Clews and Cochran, 1948). The space group is the centrosymmetric one, $P2_1/a$, with $a = 16.45$ Å, $b = 3.84$ Å, $c = 10.28$ Å and $\beta = 108°$. For this monoclinic space group there are 2_1 axes along **b** and an *a*-glide plane perpendicular to **b**. Because the *b* axis is short the structure shows up very well in projection but when the structure is projected down **b** the *a*-glide has the effect of producing a cell in which the projected contents from $x = 0$ to $a = 0.5$ are repeated in the range $x = 0.5 - 1.0$. In projection the cell is effectively halved in the *a* direction and the Patterson function corresponding to the halved cell is shown in fig. 2.4(a). The projected cell has a two-

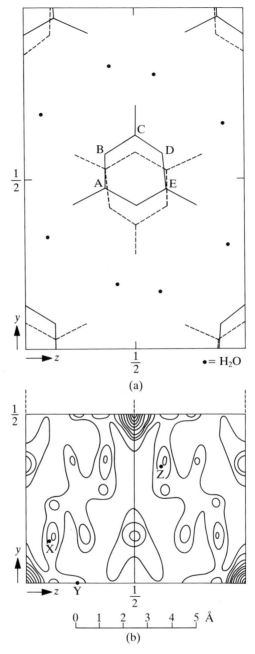

(a)

(b)

Fig. 2.3(a) The arrangement of molecules in the *a*-axis projection of 4,6-dimethyl-2-hydroxypyrimidine. (b) The Patterson peak marked **X** corresponds to the repeated vector AC (Pitt, 1948).

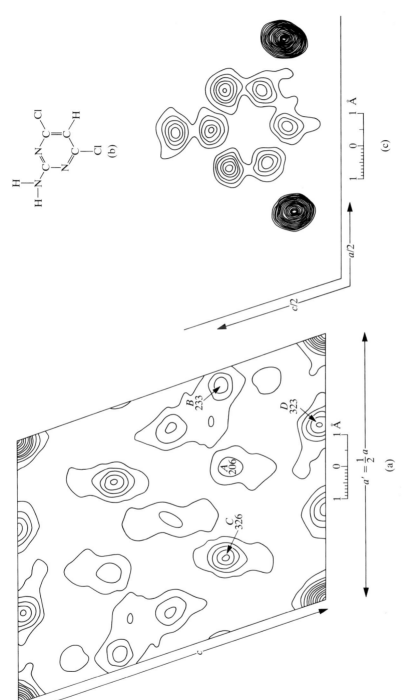

Fig. 2.4(a) The *b*-axis Patterson projection for 2-amino-4,6-dichloro-pyrimidine. (b) A molecule of 2-amino-4,6-dichloropyrimidine. (c) The electron-density map for the *b*-axis projection of 2-amino-4,6-dichloropyrimidine.

dimensional space group p2; in the projection the twofold axis is seen as a centre of symmetry.

What is needed is to find in fig. 2.4(a) the vectors corresponding to two independent chlorine atoms and their centrosymmetric partners. Let their coordinates be $\mathbf{r}_1$, $-\mathbf{r}_1$, $\mathbf{r}_2$ and $-\mathbf{r}_2$. The sixteen vectors linking them are

$$
\begin{array}{cccc}
\mathbf{0} & 2\mathbf{r}_1 & \mathbf{r}_1-\mathbf{r}_2 & \mathbf{r}_2+\mathbf{r}_1 \\
-2\mathbf{r}_1 & \mathbf{0} & -\mathbf{r}_1-\mathbf{r}_2 & -\mathbf{r}_1+\mathbf{r}_2 \\
\mathbf{r}_2-\mathbf{r}_1 & \mathbf{r}_2+\mathbf{r}_1 & \mathbf{0} & 2\mathbf{r}_2 \\
-\mathbf{r}_2-\mathbf{r}_1 & -\mathbf{r}_2+\mathbf{r}_1 & -2\mathbf{r}_2 & \mathbf{0}
\end{array} \quad . \quad (2.8)
$$

The null origin peaks do not contain useful information. The others may be summarised as single-weight peaks at $\pm 2\mathbf{r}_1$ and $\pm 2\mathbf{r}_2$ and double-weight peaks at $\pm(\mathbf{r}_1+\mathbf{r}_2)$ and $\pm(\mathbf{r}_1-\mathbf{r}_2)$.

The largest peaks in the Patterson map are those shown as A, B, C and D in fig. 2.4(a), together with their centrosymmetric counterparts. The highest peaks are C and D with coordinates in the halved cell (0.288, 0.321) and (0.665, 0.015). If these are interpreted as double-weight peaks then

$$
x_1+x_2=0.288, \qquad z_1+z_2=0.321,
$$
$$
x_1-x_2=0.665, \qquad z_1-z_2=0.015,
$$

which can be solved to give

$$
(x_1,z_1)=(0.477, 0.168), \qquad (x_2,z_2)=(-0.189, 0.153).
$$

If this interpretation is correct then we expect to find two single-weight peaks at $(2x_1,2z_1)=(0.974, 0.336)$, which is within peak B, and at $(2x_2,2z_2)=(-0.378,0.306)$. In coordinate terms, because of the periodicity in the structure, $-0.378 \equiv 0.622$ and the second vector may be identified with peak A. In this way the projected positions of the two chlorine atoms are easily found from the Patterson map.

2.1.3 Space-group-dependent vectors

In determining the positions of the chlorine atoms in the structure of 2-amino-4,6-dichloropyrimidine use was made of the interrelationships of Patterson peak positions due to the presence of the symmetry element $\bar{1}$. It was shown by Harker (1936) that for some space groups particular sections of the Patterson map show symmetry-element-dependent vectors, which can be a powerful aid to structure solution.

As a simple example consider the space group $P2_1$, which was described in

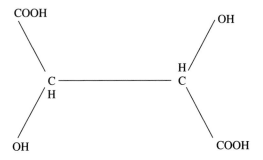

Fig. 2.5 The molecule of tartaric acid.

§1.6.3. With the 2_1 axis along **c**, atoms are related in pairs with coordinates (x, y, z) and $(\bar{x}, \bar{y}, \frac{1}{2} + z)$. The crystallographic convention of writing $\bar{x}$ instead of $-x$ should be noted here as it will be used repeatedly (see also §1.5.2). The vector between this pair is at $(2x, 2y, \frac{1}{2})$ and so the section $z = \frac{1}{2}$ shows the vectors between all pairs of atoms related by the 2_1 axis. There will also be other vectors that fall in this section just by chance but this section, usually called a *Harker section*, will be rich in information.

An excellent example of the use of the Patterson map, which included a Harker section, was given by Stern and Beevers (1950) in solving the structure of tartaric acid, shown in fig. 2.5. The space group is $P2_1$ with $a = 7.72$ Å, $b = 6.00$ Å, $c = 6.20$ Å and $\beta = 100° 10'$ with one molecule in the asymmetric unit. Atoms occur in pairs, related by the set of 2_1 axes, with coordinates of the form

$$(x, y, z); \ (\bar{x}, \tfrac{1}{2} + y, \bar{z})$$

which are linked by the vector $(2x, \frac{1}{2}, 2z)$, which will lie on the Harker section $y = \frac{1}{2}$.

An initial examination of the Harker section was not very revealing but Stern and Beevers found that the $y = 0$ Patterson section was richly populated with peaks and they interpreted the features of this section as due to a group of atoms $C_2C_3O_1O_4O_5$ lying in a plane perpendicular to **b**. This group, in the correct orientation, together with the resultant vectors in the section, $y = 0$, are shown in fig. 2.6(a). It will be seen that the general fit of vectors to peaks is quite good but it also shows that one must not expect each vector to fall on a well-shaped individual peak.

If there are two atoms with the same y coordinate at (x_1, y_0, z_1) and (x_2, y_0, z_2) then there will also be atoms at $(\bar{x}_1, \frac{1}{2} + y_0, \bar{z}_1)$ and $(\bar{x}_2, \frac{1}{2} + y_0, \bar{z}_2)$. The resultant vectors, excluding those at the origin, are single-weight Harker peaks at $\pm (2x_1, \frac{1}{2}, 2z_1)$ and $\pm (2x_2, \frac{1}{2}, 2z_2)$; and double-weight peaks

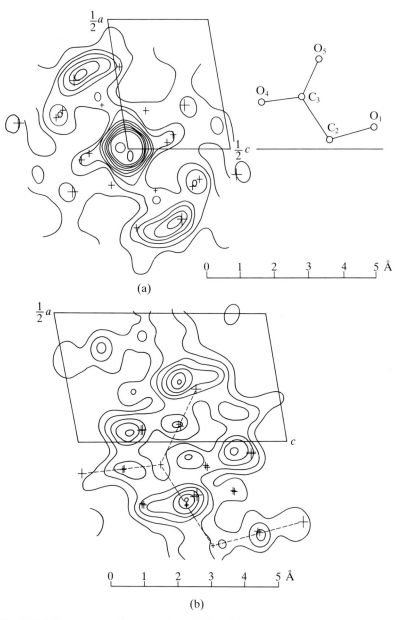

(a)

(b)

Fig. 2.6(a) Orientation of one of the $=$C.OH.COOH groups of the tartaric acid molecule compared with the central region of the $y = 0$ Patterson section. Peaks due to atoms in this group are marked by crosses, the sizes of the crosses indicating the relative weights of the peaks. (b) Patterson section at $y = \frac{1}{2}$ together with the Harker peaks due to $C_2O_1C_3O_4O_5$, marked by crosses as in (a) with double crosses indicating double-weight peaks. To appreciate the quality of the agreement all crosses should be plotted in one asymmetric unit of the Patterson section.

at $\pm(x_1-x_2,0,z_1-z_2)$ and $\pm(x_1+x_2,\frac{1}{2},z_1+z_2)$. When the Harker section was re-examined Stern and Beevers were able to look for normal Harker peaks at $\pm(2x_1,\frac{1}{2},2z_1)$ and $\pm(2x_2,\frac{1}{2},2z_2)$ together with double-weight peaks situated half way between them at $\pm(x_1+x_2,\frac{1}{2},z_1+z_2)$. In fig. 2.6(b) there is shown an interpretation of the Harker section for the group $C_2C_3O_1O_4O_5$ showing the single-weight peaks and the ten double-weight peaks that bisect the lines joining them. To show the arrangement clearly the map is shown extending outside one asymmetric unit of the Harker section. If all the peaks are plotted within the asymmetric unit outlined, a better appreciation of the quality of agreement between vectors and peaks will be obtained.

From their knowledge of the chemistry of tartaric acid Stern and Beevers were able to derive information about the remainder of the structure from other Patterson sections and so to deduce a complete trial structure for subsequent refinement.

The very common space group $P2_12_12_1$ with its three sets of twofold screw axes offers good opportunities for exploiting Harker sections. Atoms occur in sets of four with coordinates

$$(x,y,z); \; (\tfrac{1}{2}-x,\bar{y},\tfrac{1}{2}+z); \; (\tfrac{1}{2}+x,\tfrac{1}{2}-y,\bar{z}); \; (\bar{x},\tfrac{1}{2}+y,\tfrac{1}{2}-z),$$

which, apart from the null vectors between each atom and itself, gives rise to the twelve vectors

$$\begin{aligned}
&\pm(\tfrac{1}{2},\tfrac{1}{2}-2y,-2z), &&\pm(\tfrac{1}{2},\tfrac{1}{2}-2y,2z),\\
&\pm(-2x,\tfrac{1}{2},\tfrac{1}{2}-2z), &&\pm(2x,\tfrac{1}{2},\tfrac{1}{2}-2z),\\
&\pm(\tfrac{1}{2}-2x,-2y,\tfrac{1}{2}), &&\pm(\tfrac{1}{2}-2x,2y,\tfrac{1}{2}).
\end{aligned} \tag{2.9}$$

As can be seen, these fall in the Harker sections $x=\frac{1}{2}$, $y=\frac{1}{2}$ and $z=\frac{1}{2}$ but the coordinates in these sections must be self-consistent. If the positions of the major peaks in each of the Harker sections are listed then a search can be made for self-consistent triplets of peaks, one from each section. In the search allowance has to be made for slight lack of self-consistency since the positions of peaks cannot be defined with great precision. For example, let us assume that three Harker peaks have been found at $(\frac{1}{2}, 0.44, 0.91)$, $(0.46, \frac{1}{2}, 0.40)$ and $(0.89, 0.96, \frac{1}{2})$. It has been noticed that the difference of the two x coordinates is $0.89-0.46\simeq0.5$ and the same is nearly true for the differences of the y and z coordinates, 0.52 and 0.51, respectively. A solution is sought that interprets these as the three peaks

$$(\tfrac{1}{2},\tfrac{1}{2}-2y,-2z), \quad (-2x,\tfrac{1}{2},\tfrac{1}{2}-2z), \quad (\tfrac{1}{2}-2x,-2y,\tfrac{1}{2})$$

giving

$$-2x \simeq 0.46, \qquad \tfrac{1}{2}-2y \simeq 0.44, \qquad -2z \simeq 0.91,$$
$$\tfrac{1}{2}-2x \simeq 0.89, \qquad -2y \simeq 0.96, \qquad \tfrac{1}{2}-2z \simeq 0.40.$$

An approximate solution to these equations is

$$x = 0.79, \ y = 0.02_5, \ z = 0.55.$$

By interpreting the peaks in other ways it may be possible to find other values of x, y and z but these should just correspond to other choices of origin of the unit cell. In this respect it should be pointed out that the equation $2x = 0.4$ has one solution $x = 0.2$ but that $x = 0.7$ is also acceptable since, with a periodic structure, one cannot distinguish a displacement $0.4\mathbf{a}$ from a displacement $1.4\mathbf{a}$.

2.1.4 Overlap methods

It was shown by Wrinch (1939) that if a complete set of vectors between atoms were available then it would be possible to derive from it an image of the structure. We illustrate the way that this comes about in fig. 2.7. There is shown in fig. 2.7(a) a simple structure with five atoms and in fig. 2.7(b) the corresponding vector set. Some of the vectors have double weight because of the parallel bonds in the molecule but there are some single-weight peaks as well. In fig. 2.7(c) there is shown the result of superimposing the origin of a copy of the vector set on a single-weight peak of the original vector set, the two sets being kept parallel. There is an overlap of some peaks of the two vector sets and it will be seen that these show an image of the structure but in addition a centrosymmetric image. It is the intrinsic centrosymmetric nature of the vector set, and hence of the Patterson map, which leads to the pair of images.

If the original structure is centrosymmetric, as shown in fig. 2.8(a) then the vector set, given in fig. 2.8(b) has both single- and double-weight peaks. The superposition of two maps with the origin of one on a single-weight peak of the other gives a single image of the structure (fig. 2.8(c)); superposition on a double-weight peak gives a pair of images with one bonded pair of atoms in common (fig. 2.8(d)).

Although in principle a Patterson map shows the complete vector set, in practice the overlap of vectors is so severe that the individual vectors cannot be recovered. Nevertheless there have been developed some very powerful superposition techniques which are all based on locating some peak or peaks in the Patterson map that can be related to a known structural

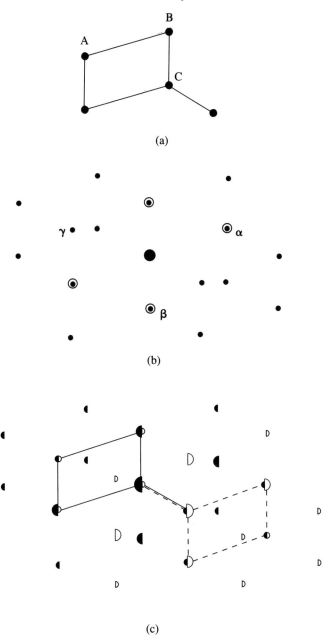

Fig. 2.7(a) A non-centrosymmetric set of points. (b) The vector set; ⊙ are double-weight peaks. (c) The superposition on a single-weight peak. For one of the overlapping vector sets: ◖ origin peak, ◖ single peak, ◖ double peak; for the second set ◗ origin peak, D single peak, ◗ double peak.

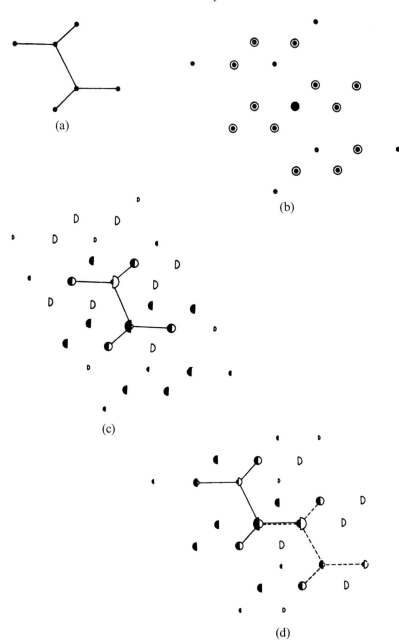

Fig. 2.8(a) A centrosymmetric set of points. (b) The vector set. (c) Superposition on a single-weight vector. (d) Superposition on a double-weight vector. (symbols as in fig. 2.7)

feature. For example, if it were known that the structure in fig. 2.7 had an atomic group within it with the form ABC then one could search for the orientation of three related vectors $\overrightarrow{AB}$, $\overrightarrow{BC}$ and $\overrightarrow{CA}$. These are shown in the vector set as α, β and γ. Now a triple superposition map is constructed with displacements $\overrightarrow{AB}$ and $\overrightarrow{CA}$ with respect to one of the maps. The reader may confirm with copies of fig. 2.7(b) on tracing paper that the triple overlap of peaks of the three superimposed maps now picks out a single image of the structure and eliminates the centrosymmetric image.

There are a whole range of image-seeking methods. Typically the starting point is a group of atoms, the relative positions of which are known in some way or perhaps even the position of a single atom. The basic requirement is that at least one vector must be known to start the image seeking process so when we say that the position of a single atom is known we assume that there are other atoms related to it by symmetry to form a group with known interatomic vectors. Let there be n atoms in the group and relative to one of them the others have positions $\mathbf{d}_1, \mathbf{d}_2 \ldots \mathbf{d}_{n-1}$. Then a superposition of n Patterson maps is made and some function of the n overlapped densities from the various maps $P(\mathbf{r})$, $P(\mathbf{r} - \mathbf{d}_1)$, $P(\mathbf{r} - \mathbf{d}_2) \ldots P(\mathbf{r} - \mathbf{d}_{n-1})$ is found. This function should have the ideal characteristic that whenever the overlapped n densities correspond to the true position of an atom then the value of the function is high. On the other hand, if the overlap point does not correspond to an atomic position then the value of the function should be low – ideally zero.

One of the problems in these types of method is that false peaks very often arise, not just due to lack of resolution of the Patterson map but sometimes also due to special relationships between interatomic vectors. This is illustrated for the simple structure shown in fig. 2.9(a) with the vector set shown in fig. 2.9(b). It is assumed that the group ABC is known and a superposition of three parallel maps is made such that their origins reproduce the shape ABC. In fig. 2.9(c) there is shown the effect of the overlap where each point corresponds to at least a single-weight peak in each of the overlapped maps. A set of peaks corresponding to the original molecule can be picked out but there are other peaks as well. We can see how this happens in fig. 2.10. If an atom D exists in the structure then the overlapped Patterson maps will contain vectors $\overrightarrow{AD}$, $\overrightarrow{BD}$ and $\overrightarrow{CD}$ that coincide. These vectors satisfy the conditions

$$\overrightarrow{AD} - \overrightarrow{BD} = \overrightarrow{AB},$$
$$\overrightarrow{BD} - \overrightarrow{CD} = \overrightarrow{BC}$$

and hence
$$\overrightarrow{CD} - \overrightarrow{AD} = \overrightarrow{CA}. \tag{2.10}$$

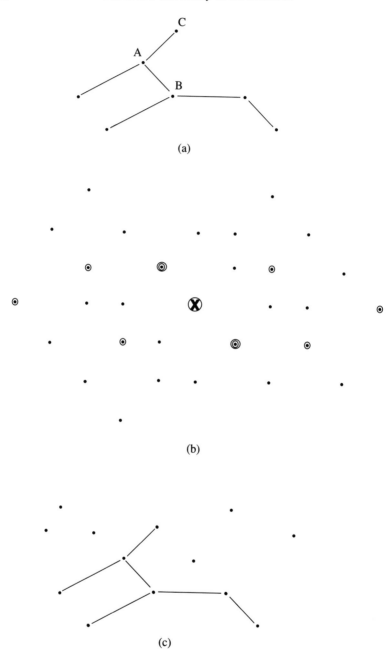

Fig. 2.9(a) The structure. (b) The vector set: ⊙ indicates a double-weight peak and ◎ a triple-weight peak. (c) Superposition based on ABC.

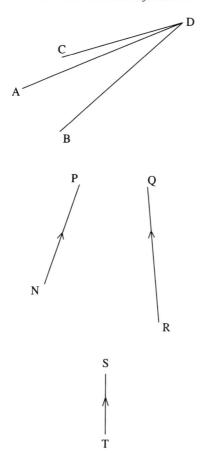

Fig. 2.10 Vectors CD, AD and BD leading to an image of atom D. The vectors $\overrightarrow{NP}$, $\overrightarrow{QR}$ and $\overrightarrow{ST}$ give a false-atom overlap.

If there happen to be three completely unrelated vectors corresponding to other parts of the structure such that

$$\overrightarrow{NP} - \overrightarrow{RQ} = \overrightarrow{AB},$$
$$\overrightarrow{RQ} - \overrightarrow{TS} = \overrightarrow{BC}$$

and hence

$$\overrightarrow{TS} - \overrightarrow{NP} = \overrightarrow{CA}. \tag{2.11}$$

then the three vectors $\overrightarrow{NP}$, $\overrightarrow{RQ}$ and $\overrightarrow{TS}$ will form an overlap to give a false peak. This is what has happened in fig. 2.9(c).

The form of the overlap function can also affect the quality of the result. The set of points in fig. 2.9(c) indicates positions at which there is at least a single-weight peak in each of the overlapped maps. However, sometimes the overlapped peaks are all of single weight and sometimes one or more of the maps give a multi-weight peak at the appropriate position. Three kinds of overlap function have been used. The first of these is the sum function

$$S(\mathbf{r}) = \sum_{i=1}^{n} P(\mathbf{r} - \mathbf{d}_i) \tag{2.12}$$

where there are n overlapped images and one of the $\mathbf{d}$ vectors can be null. The advantage of the sum function is that it can be evaluated analytically. It can be shown that a Patterson map with its origin displaced to position $-\mathbf{d}_i$ has an $\mathbf{h}^{\text{th}}$ Fourier coefficient $|F(\mathbf{h})|^2 \exp(-2\pi i \mathbf{h} \cdot \mathbf{d}_i)$ so that

$$S(\mathbf{r}) = \sum_{\mathbf{h}} |F(\mathbf{h})|^2 \left(\sum_{i=1}^{n} \exp(2\pi i \mathbf{h} \cdot \mathbf{d}_i) \right) \cos(2\pi \mathbf{h} \cdot \mathbf{r}). \tag{2.13}$$

The disadvantage of a sum function is that it is very noisy; the final map will show all the density in the overlapped maps. If at some point most of the overlapped maps show density but one or two of them do not then $S(\mathbf{r})$ will show density and if some of the maps have large spurious peaks at the point in question then the density may be as high as or even higher than that at a true atomic position.

Another function that has been advocated is the product function

$$P(\mathbf{r}) = \prod_{i=1}^{n} P(\mathbf{r} - \mathbf{d}_i). \tag{2.14}$$

There is no simple analytical form for $P(\mathbf{r})$ but it does have the advantage that if one of the maps is zero at any point then $P(\mathbf{r})$ is automatically zero at that point and any false features of the other overlapped maps will be thereby removed. A product-function map is much cleaner than a sum-function map although if some of the overlapped peaks have more than single weight then the weights of product-function peaks might badly represent the relative weights of the associated atoms.

Without doubt the most successful overlap function is the minimum function first introduced by Buerger (1950) defined as

$$M(\mathbf{r}) = \text{Min}\left[\{P(\mathbf{r} - \mathbf{d}_i)\}_{i=1}^{n}\right] \tag{2.15}$$

where the notation on the right hand side indicates the minimum value of the overlapped maps at each overlap point. This excludes a great deal of false detail. In principle if there is no density on one of the overlapped maps then there should be no atom at the corresponding point; both the product function and the minimum function achieve this end. On the other hand if the point does correspond to an atomic position and if some of the maps have multiple peaks then, whereas the product function would give greatly enhanced value at the point in question, the minimum function will equal the lowest density which, usually, would correspond to a single weight peak.

As an example we take the mineral structure Ca-seidozerite $Na_2(Na, Ca)_2(Ca, Mn)TiZr_2O_2(Si_2O_7)(F, OH)_2$ with centrosymmetric space group P2/c, $a = 5.69$ Å, $b = 7.10$ Å, $c = 18.36$ Å, $\beta = 102.7°$. The two-dimensional Patterson function of the b-axis projection is shown in fig. 2.11(a). Because of the c-glide the c axis is effectively halved in projection and, because of the centrosymmetric nature of the Patterson function, it is only necessary to show the map from $x = 0$ to $a/2$.

The vector between the centrosymmetrically related Zr atoms, the heaviest in the structure, was identified and is the one marked with a cross and this was used as the displacement vector to produce the minimum-function map in fig. 2.11(b). Various stages of refinement then lead to the final map for the projection, which is shown in fig. 2.11(c). This is a very simple example but has the merit that it is easy to interpret the result.

Minimum-function methods can be used to resolve quite complex structures if there is a group of several atoms of known orientation. A multiple overlap in which the origins have the configuration of the known group will rarely yield a complete structure but will usually indicate new structural information which can be added to the original known group. As we shall see later (§2.2) if an appreciable fraction of a structure is known then this is usually sufficient to give the structure solution.

In the language of Fourier transform theory the Patterson function is the convolution of the electron density with its own inverse or

$$P(\mathbf{r}) = \rho(\mathbf{r}) * \rho(-\mathbf{r}). \tag{2.16}$$

From the convolution theorem this gives the Fourier transform of $P(\mathbf{r})$ as the product of the Fourier transforms of $\rho(\mathbf{r})$ and $\rho(-\mathbf{r})$ or

$$FT\{P(\mathbf{r})\} = F(\mathbf{h}) \times F(\bar{\mathbf{h}}) = |F(\mathbf{h})|^2 \tag{2.17}$$

which, apart from the scaling constant $1/V$, is the result indicated in equations (2.6) and (2.7).

This interpretation as a convolution reveals that the Patterson map of a structure containing N atoms in the unit cell will show N parallel images of the structure plus N inverted images. Of course when N is large then, because of heavy overlap, the individual images cannot be distinguished

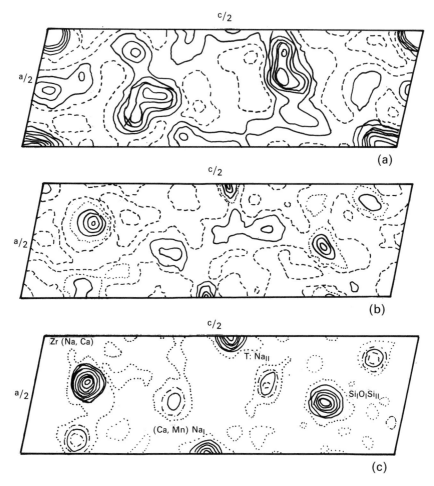

Fig. 2.11(a) The Patterson function for the *b*-axis projection of Ca-seidozerite. The vector between centrosymmetrically related Zr atoms is shown by X. (b) A minimum function based on the Zr–Zr vector. (c) The finally refined structure.

but, as we have already seen, for a non-centrosymmetric structure a minimum function based on a single-weight peak should, in principle, separate out one image of a structure plus a centrosymmetric image. The two difficulties with this simple idea are that, firstly, there would normally be so much noise in the minimum-function map that it would be impossible to identify the image and, secondly, that single weight peaks tend to be swamped by overlap and fluctuations due to error in the observed data. While various ideas have been advanced whereby, in principle, the Patterson function could be deconvoluted to reveal the structure, e.g. Germain and Woolfson (1966), in practice this cannot be done for structures of interest.

2.1.5 Rotation and translation functions

Very often the form of the molecule whose crystal structure is being sought will be known either completely or in part. In such a case, for a particular orientation, we know the relative coordinates of a set of M atoms, $(\mathbf{r}_i)_{rel}$ where i ranges from 1 to M, and the problem of fixing this unit in the unit cell consists of finding rotation and translation matrices R and t such that the true coordinates in the unit cell are found from

$$\mathbf{r}_i = R(\mathbf{r}_i)_{rel} + t. \tag{2.18}$$

The rotation matrix can be defined in terms of three Euler angles (α, β, γ), which define the orientation of the group and the translation matrix is defined by three displacements along cell-edge directions (X, Y, Z). The rotation matrix will be of the form

$$R = \begin{bmatrix} \cos\alpha\cos\beta\cos\gamma & -\sin\alpha\cos\beta\cos\gamma & -\sin\beta\cos\gamma \\ -\sin\alpha\sin\gamma & +\cos\alpha\sin\gamma & \\ -\cos\alpha\cos\beta\sin\gamma & -\sin\alpha\cos\beta\sin\gamma & \sin\beta\sin\gamma \\ -\sin\alpha\cos\gamma & +\cos\alpha\cos\gamma & \\ \cos\alpha\sin\beta & \sin\alpha\sin\beta & \cos\beta \end{bmatrix}. \tag{2.19}$$

It is possible to divide the problem of finding the rotation and translation into two separate parts. If the values of $\mathbf{r}_i$ are found for a particular set of values of (α, β, γ) then coincidence can be sought between the vectors $\mathbf{r}_i - \mathbf{r}_j (i, j = 1 - M)$ and peaks in the Patterson map. A function defining the coincidence can be calculated for all (α, β, γ) and plotted on a map such that the highest peak indicates the best fit. Once the orientation is fixed then the vectors between the atoms in symmetry-related groups can be found as functions of (X, Y, Z); by plotting a map describing the coincidence of these vectors with Patterson peaks the correct translation can usually be found. Various functions have been advanced to describe the fit of vectors to the Patterson map and these resemble the sum, product and minimum functions defined in equations (2.14) to (2.16). These are

$$\mathscr{S}(\alpha, \beta, \gamma) = \sum_{i=1}^{M} \sum_{j=1}^{M} P(\mathbf{r}_i - \mathbf{r}_j), \tag{2.20}$$

$$\mathscr{P}(\alpha,\beta,\gamma) = \prod_{i,j=1}^{M} P(\mathbf{r}_i - \mathbf{r}_j), \tag{2.21}$$

$$\mathscr{M}(\alpha,\beta,\gamma) = \mathrm{Min}\left[\{P(\mathbf{r}_i - \mathbf{r}_j)\}_{i,j=1}^{M}\right]. \tag{2.22}$$

Other functions have also been advocated from time to time, e.g. Nordman (1966), but the general principle, that of matching a set of vectors to peaks in a map is the same. Tollin and Cochran (1964) have made use of the sum function and they have also transformed the problem into one in reciprocal space. For each set of angles (α,β,γ) the coordinates $\mathbf{r}_i$ are calculated and then the sum function (2.20) is evaluated as follows. From equation (2.2)

$$P(\mathbf{r}_i - \mathbf{r}_j) = \frac{1}{V}\sum_{\mathbf{h}}|F(\mathbf{h})|^2 \cos\left[2\pi\mathbf{h}\cdot(\mathbf{r}_i - \mathbf{r}_j)\right]$$

$$= \frac{1}{V}\sum_{\mathbf{h}}|F(\mathbf{h})|^2\left[\cos(2\pi\mathbf{h}\cdot\mathbf{r}_i)\cos(2\pi\mathbf{h}\cdot\mathbf{r}_j) + \sin(2\pi\mathbf{h}\cdot\mathbf{r}_i)\sin(2\pi\mathbf{h}\cdot\mathbf{r}_j)\right]. \tag{2.23}$$

The sum function can now be written

$$\mathscr{S}(\alpha,\beta,\gamma) = \frac{1}{V}\sum_{\mathbf{h}}|F(\mathbf{h})|^2\left\{\left[\sum_{i=1}^{M}\cos(2\pi\mathbf{h}\cdot\mathbf{r}_i)\right]^2 + \left[\sum_{i=1}^{M}\sin(2\pi\mathbf{h}\cdot\mathbf{r}_i)\right]^2\right\}, \tag{2.24}$$

which is convenient to calculate. If the atoms in the known group are of very different atomic numbers then the cosine and sine terms in (2.24) should be given appropriate weights, Z_i.

The range of Eulerian angles that must be searched depends on the symmetry of the crystal and of the model. The logic behind this is easily visualised; since in space group $P2_12_12_1$, for example, each molecule or group occurs four times in symmetry-related positions and orientations then only one quarter of direction-space needs to be explored. This matter has been fully discussed by Tollin, Main and Rossman (1966). There are other useful modifications of the technique including the removal of very short vectors from the sum on the right-hand side of equation (2.23). Since the short vectors correspond to the very highly occupied origin-peak region of the Patterson map removing them increases the discrimination of the sum-function.

If the known group is planar then the rotation-function calculations are much simplified. The problem can be factorised into first finding two angles that define the plane of the group with respect to the cell axes followed by determining the azimuthal angle, φ, defining the orientation of the group in that plane. It is more convenient for the first part of this process to use spherical polar angles (θ, ϕ) to describe the direction of the normal to the planar group and this is done with respect to a set of orthogonal axes $\mathbf{a}^*$ (the vector corresponding to one side of the reciprocal cell), $\mathbf{b}$ (the real cell vector) and $\mathbf{a}^* \wedge \mathbf{b}$; θ is the angle between the normal to the plane and the direction $\mathbf{a}^* \wedge \mathbf{b}$.

A function $I(\theta, \phi)$ is calculated that will recognise that a plane in the Patterson map which passes through the origin contains many vectors. To link this with the planar group the Patterson map is only examined in a disc-shaped region with its centre at the origin. The function that is calculated is

$$I(\theta, \phi) = \int_V P(\mathbf{r})D(\mathbf{r})\,dV, \qquad (2.25)$$

where

$$D(\mathbf{r}) = 1 \text{ for a point on the disc,}$$
$$D(\mathbf{r}) = 0 \text{ elsewhere.}$$

Ignoring some constant factors this can be expressed in reciprocal-space terms as

$$I(\theta, \phi) = \sum_{\mathbf{h}} |F(\mathbf{h})|^2 2\pi R^2 \frac{J_1(2\pi RS)}{2\pi RS}, \qquad (2.26)$$

where $J_1(x)$ is a modified Bessel function, R is the radius of the disc and S the perpendicular distance of the reciprocal lattice point $\mathbf{h}$ from the normal to the disc. For each direction of the disc normal, (θ, ϕ), the distance S is calculated for each reciprocal lattice point $\mathbf{h}$ and then the summation is made. Not every direction needs to be explored. For a monoclinic crystal it is sufficient to explore the range $0 \le \phi \le \pi$ while for an orthorhombic crystal this is further restricted to $0 \le \phi \le \pi/2$. However, since the integrand in (2.26) does not have the symmetry of the crystal, it is necessary to include all symmetry-related reflections, other than Friedel opposites, in the summation.

A two-dimensional plot of $I(\theta, \phi)$ can conveniently be made on a Samson–Flamsteed sinusoidal equal-area projection and one of these is shown in fig. 2.12 for the purine isosine (Tollin and Cochran, 1964) where the direction of the normal to the plane is clearly indicated.

A simple way to calculate the azimuthal angle, φ, which defines the orientation in the plane is to calculate a section through the origin of the Patterson map in the plane indicated by $I(\theta, \phi)$ and then find the best fit of

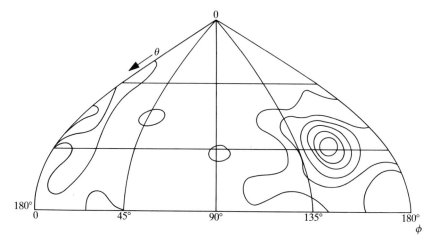

Fig. 2.12 A rotation-function map for isosine.

the vector set of the planar group. This method requires calculation of the three-dimensional coordinates corresponding to points in a grid in the plane but this is a simple exercise in geometry and will not be discussed further here.

Once the orientation of a group, planar or non-planar, has been fixed then it is possible to fix the position of the group within the unit cell. In practice, where there are many symmetry elements in the space group, it is possible to break up the problem into finding the position of the group in relation to different symmetry elements in turn. However, with the power of modern computers this should not be necessary.

The vectors that are to be matched with the Patterson map are those between atoms in symmetry-related groups. If $\mathbf{r}_i$ is the position of an atom in one group then there will be other atoms in related positions at positions $T_k\mathbf{r}_i$ where T_k is some symmetry operator. A complete set of symmetry-related positions may be found by taking a set of T_k; thus for P2$_1$2$_1$2$_1$ the appropriate set of T operators would give

$$T_1(x,y,z) = (x,y,z),$$
$$T_2(x,y,z) = (\bar{x}, \tfrac{1}{2}+y, \tfrac{1}{2}-z),$$
$$T_3(x,y,z) = (\tfrac{1}{2}-x, \bar{y}, \tfrac{1}{2}+z),$$
$$T_4(x,y,z) = (\tfrac{1}{2}+x, \tfrac{1}{2}-y, \bar{z}). \tag{2.27}$$

The group location function is now defined as

$$Q(\mathbf{R}) = \sum_k \sum_i \sum_j P[(\mathbf{r}_i + \mathbf{R}) - T_k(\mathbf{r}_j + \mathbf{R})] \tag{2.28}$$

and it is only necessary to range **R** over one asymmetric unit of the cell. This can be tranformed into a reciprocal space calculation so that

$$Q(\mathbf{R}) = \sum_{\mathbf{h}} \sum_{k} \sum_{i} \sum_{j} |F(\mathbf{h})|^2 \cos\left[(\mathbf{r}_i + \mathbf{R}) - T_k(\mathbf{r}_j + \mathbf{R})\right]. \tag{2.29}$$

A peak in this function indicates a good match of the intergroup vectors.

The Patterson function was the dominant tool for the solution of crystal structures up to about 1970 but it was thereafter overtaken by direct methods (chapter 3). There has recently been a resurgence of interest in Patterson methods, spurred on by the increasing power of computers, and there are new approaches in which it is suggested that Patterson and direct methods should be combined (e.g. Giacovazzo, 1991).

2.2 The heavy-atom method

2.2.1 Wilson statistics

It was seen in §2.1.2 that analysis of the Patterson function can sometimes reveal the locations of heavier atoms in the structure or of groups of lighter atoms with helpful symmetry. Here we shall consider how to exploit such information to reveal the whole structure. However, before doing so it is necessary to understand the statistical basis of how a small part of a structure contributes to the total structure factor. It is intuitively obvious that if almost all the structure is known then the phase contributed by that part will be very close to the true phase. A Fourier synthesis, calculated with these phases inserted in equation (1.37), is then likely to show the total structure – or at least more than the portion of the structure from which the phases were derived. Conversely, if a very tiny part of the structure is known, say a few light atoms in a structure containing several hundred atoms, then its contribution to the total structure factor will be too small to give a reliable phase indication. The situation most frequently encountered is intermediate between these two extremes and to discover how best to use the information from a known part of the structure it is necessary to examine the probability density of the magnitudes of structure factors, first derived by Wilson (1949).

Wilson's approach was to regard the right-hand-side contributions to the structure factor equation (1.21) as random variables and to apply the Central-limit Theorem to find the distribution of $|F(\mathbf{h})|$. This theorem states that the sum of a large (ideally infinite) number of independent random variables will have a normal distribution whose mean is the sum of the means of the random variables and whose variance is the sum of their variances. This may be most easily illustrated for a centrosymmetric crystal where the structure-factor equation is of the form

$$F(\mathbf{h}) = \sum_{j=1}^{N/2} 2f_j \cos(2\pi\mathbf{h}\cdot\mathbf{r}_j). \tag{2.30}$$

The reason for the difference in form between equations (1.23) and (2.30) is that the contributions of atoms at $\mathbf{r}_j$ and $-\mathbf{r}_j$ are not independent, which is a requirement of the Central-limit Theorem.

Each $2f_j \cos(2\pi\mathbf{h}\cdot\mathbf{r}_j)$, say ξ_j, is now considered as a random variable with mean

$$\overline{\xi_j} = 0,$$

and variance

$$V_j = \overline{\xi^2} - \overline{\xi}^2 = \overline{4f_j^2 \cos^2(2\pi\mathbf{h}\cdot\mathbf{r}_j)}$$
$$= 2f_j^2$$

since the mean value of $\cos^2\theta$, where θ is a random variable, is 0.5. Thus, for a centrosymmetric structure with many atoms, the values of $F(\mathbf{h})$ have a normal distribution with mean

$$\sum_{j=1}^{n/2} \overline{\xi_j} = 0$$

and variance

$$\sum_{j=1}^{N/2} 2f_j^2 = \sum_{j=1}^{N} f_j^2 = \Sigma.$$

The form of this distribution is

$$P_{\bar{1}}(F) = (2\pi\Sigma)^{-\frac{1}{2}} \exp(-F/2\Sigma). \tag{2.31a}$$

The distribution for the magnitude of F, $|F|$, is then

$$P_{\bar{1}}(|F|) = (2/\pi\Sigma)^{\frac{1}{2}} \exp(-|F|^2/2\Sigma). \tag{2.31b}$$

For a non-centrosymmetric structure equation (1.21) may be written in the form

$$F(\mathbf{h}) = \sum_{j=1}^{N} f_j \cos(2\pi\mathbf{h}\cdot\mathbf{r}_j) + i \sum_{j=1}^{N} f_j \sin(2\pi\mathbf{h}\cdot\mathbf{r}_j) = A + iB. \tag{2.32}$$

Both A and B have a distribution similar in form to (2.31a) except that, since the atoms do not occur in centrosymmetric pairs, Σ in equation (2.31a) is

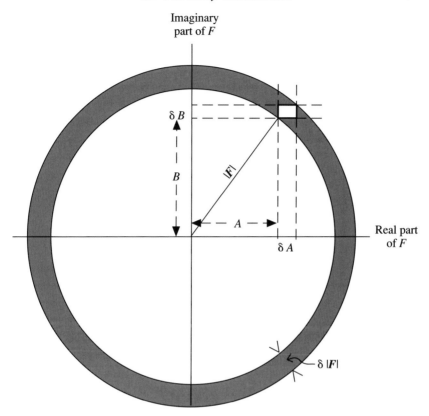

Fig. 2.13 The regions of space both between A and $A + \delta A$ and B and $B + \delta B$ (boldly-outlined rectangle) and between $|F|$ and $|F| + \delta |F|$ (shaded).

replaced by $\frac{1}{2}\Sigma$. The relationship between F, A and B is shown on an Argand diagram in fig. 2.13. The probability that the real part of the structure factor is between A and $A + \delta A$ and that the imaginary part is between B and $B + \delta B$ is the product of two independent probabilities, i.e.

$$P(A, B)\delta A \, \delta B = P(A)P(B)\delta A \, \delta B = (\pi\Sigma)^{-1}\exp[-(A^2 + B^2)/\Sigma]\delta A \, \delta B$$
$$= (\pi\Sigma)^{-1}\exp(-|F|^2/\Sigma)\,\delta A \, \delta B. \tag{2.33}$$

This probability is proportional to the area $\delta A \, \delta B$ shown in the diagram and is functionally dependent only on $|F|$. Thus the probability that the structure amplitude will be between $|F|$ and $|F| + \delta|F|$, i.e. in the shaded region shown in fig. 2.13, will be

$$P_1(|F|)\,\delta|F| = (\pi\Sigma)^{-1}\exp(-|F|^2/\Sigma) \times 2\pi|F|\,\delta|F|$$

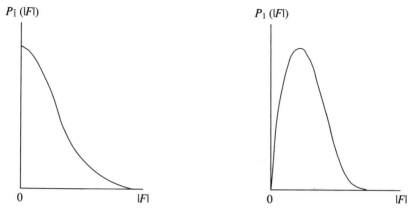

Fig. 2.14 The centric and acentric distributions $P_{\bar{1}}(|F|)$ and $P_1(|F|)$.

or

$$P_1(|F|) = (2/\Sigma)|F| \exp(-|F|^2/\Sigma). \qquad (2.34)$$

The two distributions, (2.31b) and (2.34), are shown in fig. 2.14. The differences between them can be used to distinguish between centrosymmetric and non-centrosymmetric structures. An examination of the two distributions reveals that the centric distribution (from a centrosymmetric structure) gives a greater proportion of very small structure amplitudes but also a higher proportion of large ones. The acentric distribution has a distribution more closely clustered around the value of $\overline{|F|}$.

A quantity that enables centrosymmetric and non-centrosymmetric structures to be distinguished, but does not require a knowledge of the quantity Σ, is the ratio

$$M = \frac{\overline{|F|}^2}{\overline{|F^2|}}. \qquad (2.35)$$

For the centric distribution $M = 2/\pi = 0.637$ while for the acentric distribution it is $\pi/4 = 0.785$; for both distributions $\overline{|F^2|} = \Sigma$. Since the values of $|F|$ and $\overline{|F^2|}$ depend on the scattering and temperature factors, which fall off with increasing scattering angle, it is necessary to find values of M in fairly narrow shells in reciprocal space within which these factors are sensibly constant. The values of M from different shells can then be averaged to give a global value for the complete data set.

2.2.2 The heavy-atom method; the centrosymmetric structure

We now consider a centrosymmetric structure for which the positions of a part are known – from an analysis of the Patterson function, for example – while the positions of the remaining N' atoms have to be found. Woolfson (1956) derived from theoretical considerations the proportion of correct sign indications that would be given by heavy atoms, or a group of lighter atoms, in known positions and also described how to derive maximum information about the positions of the remaining atoms.

The total structure factor may be written as

$$F(\mathbf{h}) = K(\mathbf{h}) + \sum_{j=1}^{N'} f_j \cos(2\pi \mathbf{h} \cdot \mathbf{r}_j) = K(\mathbf{h}) + F'(\mathbf{h}), \qquad (2.36)$$

where $K(\mathbf{h})$ is the contribution of the known part of the structure and $F'(\mathbf{h})$ that of the unknown part. For a centrosymmetric structure the phase is either 0 or π, which is to say that $F(\mathbf{h})$ is either positive or negative. From (2.36)

$$F'(\mathbf{h}) = F(\mathbf{h}) - K(\mathbf{h}). \qquad (2.37)$$

If $F(\mathbf{h})$ and $K(\mathbf{h})$ have the same sign then

$$|F'(\mathbf{h})| = \big||F(\mathbf{h})| - |K(\mathbf{h})|\big| \qquad (2.38a)$$

while if their signs are different then

$$|F'(\mathbf{h})| = \big||F(\mathbf{h})| + |K(\mathbf{h})|\big|. \qquad (2.38b)$$

The ratio of the probabilities of having the same sign, P_+, to having different signs, P_-, will be the ratio of having the two associated magnitudes according to the distribution (2.31b).

This gives

$$\frac{P_+}{P_-} = \frac{\exp\big[-(|F(\mathbf{h})| - |K(\mathbf{h})|)^2/2\Sigma'\big]}{\exp\big[-(|F(\mathbf{h})| + |K(\mathbf{h})|)^2/2\Sigma'\big]} = \exp(2|F(\mathbf{h})|\,|K(\mathbf{h})|/\Sigma'), \quad (2.39)$$

where

$$\Sigma' = \sum_{j=1}^{N'} f_j^2.$$

Since $P_+ + P = 1$ it follows from (2.39) that the probability that the sign of $K(\mathbf{h})$ indicates the correct sign of $F(\mathbf{h})$ is

$$P_+(\mathbf{h}) = \tfrac{1}{2} + \tfrac{1}{2}\tanh(|F(\mathbf{h})|\,|K(\mathbf{h})|/\Sigma'). \qquad (2.40)$$

Some examples will illustrate the application of (2.40). We consider a structure with 200 carbon atoms in the unit cell with four chlorine atoms in known positions. We take f_C as 2.5, giving $\Sigma' = 1250$ and f_{Cl} as 9.5. The values of P_+ for various values of $|F(\mathbf{h})|$ and $|K(\mathbf{h})|$ are indicated below.

$|F(\mathbf{h})|$	$|K(\mathbf{h})|$	P_+	$|F(\mathbf{h})|$	$|K(\mathbf{h})|$	P_+
60	10	0.723	100	10	0.832
60	20	0.872	100	20	0.961
60	35	0.966	100	35	0.996
80	10	0.782	120	10	0.872
80	20	0.928	120	20	0.979
80	35	0.989	120	35	0.999

The above values are quite realistic for the structure described and it is clear that the information from four chlorine atoms will lead to many reliable sign indications for the structure factors and hence to the solution of the structure.

Woolfson also showed that a Fourier synthesis calculated with coefficients

$$X(\mathbf{h}) = S(K)\left[|F(\mathbf{h})| \tanh\left(\frac{|F(\mathbf{h})K(\mathbf{h})|}{\Sigma'}\right) - K(\mathbf{h}) \right] \qquad (2.41)$$

where $S(K)$ is the sign of $K(\mathbf{h})$, would give a map in which the known atoms did not appear and where the density due to unknown atoms would have the highest possible signal-to-noise ratio. The advantage in removing the known atoms from the map, especially if they are heavy atoms, is that it also removes the diffraction ripples associated with them, which could distort the appearance of density corresponding to the atoms one wished to locate.

It might be thought that the greater the contribution of the known part of the structure the better is the situation for resolving the complete structure. This is true up to a certain point. If the contribution of the heavy, or known, atoms is too large then the errors in intensity measurements may be comparable to the contribution of the unknown atoms. In this case the phases may be determined well but the Fourier map may show the unknown part of the structure badly or even not at all. Experience suggests that an optimum situation is where the average contribution of the known part of the structure equals that of the unknown part, or

$$\sum_{\text{known}} f^2 = \sum_{\text{unknown}} f^2.$$

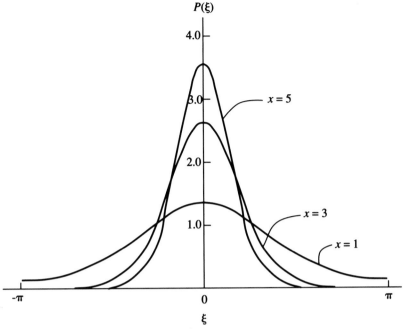

Fig. 2.15 The probability distribution $P(\xi)$.

2.2.3 The heavy-atom method; the non-centrosymmetric structure

Sim (1957; 1959; 1960) extended the theory of the heavy-atom method to non-centrosymmetric structures and the results that he obtained have been extensively applied and have become important in many different areas of structure determination.

We again consider a reflection of index **h** with structure amplitude $|F(\mathbf{h})|$ for which part of the structure is known and makes a contribution to $F(\mathbf{h})$ of

$$K(\mathbf{h}) = |K(\mathbf{h})| \exp\left[i\,\alpha(\mathbf{h})\right]. \tag{2.42}$$

If the phase of the complete structure factor $F(\mathbf{h})$ is $\phi(\mathbf{h})$ and

$$\xi = \phi(\mathbf{h}) - \alpha(\mathbf{h}) \tag{2.43}$$

then Sim (1960) showed that the probability distribution of ξ is

$$P(\xi) = \frac{\exp\left[X\cos(\xi)\right]}{2\pi I_0(X)} \tag{2.44}$$

where I_0 is the modified zeroth-order Bessel function and $X = 2|F(\mathbf{h})K(\mathbf{h})|/\Sigma'$. This distribution is shown in fig. 2.15 for various values of X; it is clear

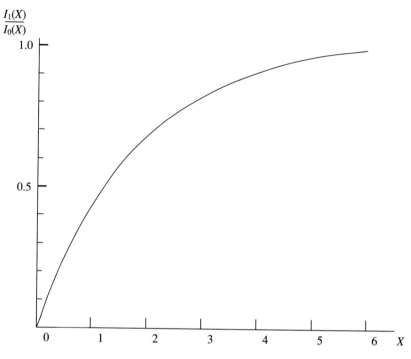

Fig. 2.16 The Sim weighting function.

that its maximum is for $\xi = 0$, that is to say that the most probable value for $\phi(\mathbf{h})$ is $\alpha(\mathbf{h})$, and the spread, or variance, of the distribution decreases as the value of X increases.

A Fourier synthesis with structure amplitudes $|F(\mathbf{h})|$ and phases $\alpha(\mathbf{h})$ would be expected to show up the positions of the unknown atoms but Sim (1960) showed that, as in the case of the centrosymmetric structure, a weighted Fourier synthesis

$$\rho'(\mathbf{r}) = \sum_{\mathbf{h}} W(\mathbf{h})|F(\mathbf{h})| \cos\left[2\pi\mathbf{h}\cdot\mathbf{r} - \alpha(\mathbf{h})\right] \qquad (2.45)$$

gives the least error in electron density. In this case the weight is

$$W(\mathbf{h}) = I_1(X)/I_0(X), \qquad (2.46)$$

where I_1 is a first-order modified Bessel function and $\rho'(\mathbf{r})$ includes the density due to the heavy, or known, atoms. The form of $I_1(X)/I_0(X)$ as a function of X is shown in fig. 2.16.

As has previously been mentioned, the interpretation of a Patterson function can lead to knowledge of only some part of the structure and this is also true of direct methods, described in the next chapter. Sim-weighted Fourier syntheses have proved to be an invaluable way of extending a partial structure to the full solution.

3
Direct methods

3.1 Introduction to direct methods

3.1.1 The basis of direct methods

The term *direct methods* is used to describe that class of methods for determining crystal structures that find estimates of the values of phases *ab initio* from the magnitudes of the structure factors. With a knowledge of the phases, or a sufficiently large subset of them, an electron-density map can be calculated with equation (1.36) to reveal the structure.

It is simple to show that the magnitudes of structure factors must constrain the phases in some way. In fig. 3.1 there is shown the projection down the *z* axis of a centrosymmetric unit cell. If the structure factors of index (6 1 0), (4 4 0) and (2 $\bar{3}$ 0) all have their maximum possible magnitudes then the atoms must all lie either on the appropriate sets of full lines (positive structure factor) or broken lines (negative structure factor). If, for example, both $F(6\ 1\ 0)$ and $F(4\ 4\ 0)$ are positive then the atoms must all lie on the intersection of the two sets of full lines for these reflections. Inspection of fig. 3.1 then shows that $F(2\ \bar{3}\ 0)$ is also positive because it is only the full lines for that reflection which pass through the intersections of the other sets of full lines. Consideration of other possibilities leads to the conclusion that whatever the individual signs of these three reflections

$$s(6\ 1\ 0)s(4\ 4\ 0)s(2\ \bar{3}\ 0) = +1 \qquad (3.1)$$

where $s(h,k,l)$ is the sign (± 1) of $F(h,k,l)$.

It may be seen that the indices of the three reflections are linked as

$$(6\ 1\ 0) = (4\ 4\ 0) + (2\ \bar{3}\ 0) \qquad (3.2)$$

and it is readily shown that, in general, if $F(\mathbf{h}_1)$, $F(\mathbf{h}_2)$ and $F(\mathbf{h}_1 + \mathbf{h}_2)$ all have maximum magnitudes then

$(2, \bar{3}, 0)$　　　$(4, 4, 0)$　　　　　　　　　　　$(6, 1, 0)$

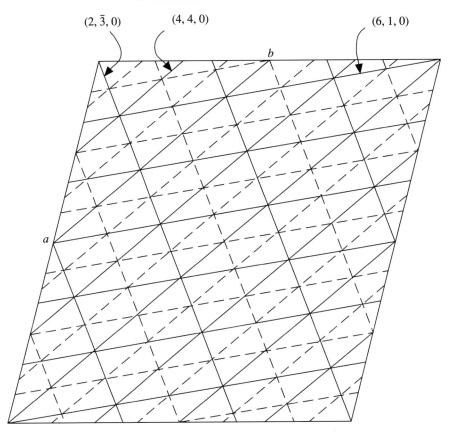

Fig. 3.1 Full lines give locations of $\cos(2\pi\mathbf{h}\cdot\mathbf{r}) = +1$ and dashed lines locations of $\cos(2\pi\mathbf{h}\cdot\mathbf{r}) = -1$ for the indices shown. All triple intersections correspond to a product of values equal to $+1$.

$$s(\mathbf{h}_1 + \mathbf{h}_2) = s(\mathbf{h}_1)s(\mathbf{h}_2). \tag{3.3}$$

For a centrosymmetric structure we may write

$$F(\mathbf{h}_1 + \mathbf{h}_2) = \sum_{j=1}^{N} f_j \cos\left[2\pi(\mathbf{h}_1 + \mathbf{h}_2)\right]\cdot\mathbf{r}_j$$

$$= \sum_{j=1}^{N} f_j \cos(2\pi\mathbf{h}_1\cdot\mathbf{r}_j)\cos(2\pi\mathbf{h}_2\cdot\mathbf{r}_j) \tag{3.4}$$

since, with $\cos(2\pi\mathbf{h}_1\cdot\mathbf{r}_j)$ and $\cos(2\pi\mathbf{h}_2\cdot\mathbf{r}_j)$ at extremum values, $\sin(2\pi\mathbf{h}_1\cdot\mathbf{r}_j)$ and $\sin(2\pi\mathbf{h}_2\cdot\mathbf{r}_j)$ must be zero for all j. In the case we are taking, the value of $\cos(2\pi\mathbf{h}_1\cdot\mathbf{r}_j)$ is independent of j and equal to either $+1$ or -1 and the same is true for $\cos(2\pi\mathbf{h}_2\cdot\mathbf{r}_j)$. It will be seen from this that equation (3.3) follows from (3.4). While the above discussion and analysis have depended on having structure amplitudes with maximum possible values it will be intuitively obvious that even if the amplitudes are somewhat less than the maximum possible some constraints may still apply.

3.1.2 Unitary and normalised structure factors

The maximum magnitudes of structure factors will differ because scattering factors, which depend on scattering angle and intensities, from which experimentally derived structure factors are found, will be affected by different Debye–Waller factors. From a theoretical point of view it is useful to have structure factors normalised in some way to remove scattering-angle-dependent effects.

In early work on direct methods a modified structure factor called the *unitary structure factor* was used. Starting from the structure-factor equation

$$F(\mathbf{h}) = \sum_{j=1}^{N} f_j \exp(2\pi i \mathbf{h}\cdot\mathbf{r}_j) \tag{3.5}$$

both sides are divided by $\sum_{j=1}^{N} f_j$ to give

$$U(\mathbf{h}) = F(\mathbf{h}) \bigg/ \left|\sum_{j=1}^{N} f_j\right| = \sum_{j=1}^{N} n_j \exp(2\pi i \mathbf{h}\cdot\mathbf{r}_j), \tag{3.6}$$

where $U(\mathbf{h})$ is the unitary structure factor and

$$n_j = f_j \bigg/ \left|\sum_{j=1}^{N} f_j\right|$$

is the *unitary scattering factor*. The values of n_j satisfy the equation

$$\sum_{j=1}^{N} n_j = 1. \tag{3.7}$$

Since the maximum magnitude of the exponential terms in (3.5) is unity it follows that

$$|F(\mathbf{h})| \le \sum_{j=1}^{N} f_j$$

so that

$$|U(\mathbf{h})| \le 1. \tag{3.8}$$

Of greater importance in the application of direct methods is the *normalised structure factor* introduced by Hauptman and Karle (1953). This is defined by

$$E(\mathbf{h}) = F(\mathbf{h})/(\varepsilon\Sigma)^{\frac{1}{2}}. \tag{3.9}$$

Σ is the quantity used in equation (2.31a) and is the average value of $|F|^2$ for the scattering angle corresponding to $\mathbf{h}$. The factor ε usually equals unity but is, in general, a small integer that depends on the space group and the type of reflection and gives the effect that for all the reflections of that particular type $\overline{|E|^2} = 1$. To illustrate this, if a space group has a mirror plane perpendicular to b then, in projection down the y axis, pairs of similar atoms project on top of each other. For a reflection of type $(h, 0, l)$ the structure is equivalent to having half the number of atoms ($N' = N/2$) but each of twice the scattering power ($f_j' = 2f_j$). The average $|F|^2$ for such reflections at any particular scattering angle will be

$$\Sigma' = \sum_{j=1}^{N/2} (2f_j)^2 = 2\Sigma. \tag{3.10}$$

It is the square root of this that is the divisor in (3.9) so that $\varepsilon = 2$ in this case.

A theoretical advantage of normalised structure factors is that the statistical distribution of their amplitudes depends only on the space group and not on the complexity of the structure. The Wilson distributions for $|E|$ for space groups $P\bar{1}$ and $P1$ are

$$P_{\bar{1}}(|E|) = (2/\pi)^{\frac{1}{2}} \exp(-|E|^2/2), \tag{3.11}$$

$$P_1(|E|) = 2|E| \exp(-|E|^2). \tag{3.12}$$

The root-mean-square value of $|E|$ is unity so that, for example, there will be very few values of $|E| > 3$ for normal structures.

Values of E and U are related so that for a general reflection ($\varepsilon = 1$) and an equal-atom structure

$$E = N^{\frac{1}{2}} \times U \tag{3.13}$$

where N is the number of atoms in the unit cell.

3.2 Inequality relationships

3.2.1 Harker–Kasper inequalities

Modern direct methods began with the derivation of inequality relationships by Harker and Kasper (1948). These are best described in terms of the unitary structure factors, as given in (3.6), and the assumption is made that the value of n_j is a constant throughout reciprocal space. This assumption is not strictly true since the scattering factor curves do not all have the same relative fall-off (see fig. 1.9); for $(\sin\theta)/\lambda = 0$ the ratio of scattering factors of sulphur and carbon is 2.667 while for $(\sin\theta)/\lambda = 0.5$ the ratio is 4.243.

The mathematical background used by Harker and Kasper was the Cauchy inequality

$$\left|\sum_{j=1}^{N} a_j b_j\right|^2 \le \left(\sum_{j=1}^{N} |a_j|^2\right)\left(\sum_{j=1}^{N} |b_j|^2\right) \tag{3.14}$$

where the a and b terms can be real or complex quantities.

We now apply (3.14) to the unitary structure factor equation

$$U(\mathbf{h}) = \sum_{j=1}^{N} n_j \exp(2\pi i \mathbf{h} \cdot \mathbf{r}_j)$$

partitioning the term in the summation as $a_j = n_j^{1/2}$ and $b_j = n_j^{1/2} \exp(2\pi i \mathbf{h} \cdot \mathbf{r}_j)$.

If n_j is real, which it is for normal (as distinct from anomalous) X-ray scattering, then $|a_j|^2 = n_j$ and $|b_j|^2 = n_j$.

Since, from (3.7)

$$\sum_{j=1}^{N} |a_j|^2 = \sum_{j=1}^{N} |b_j|^2 = 1$$

the Cauchy inequality gives

$$|U(\mathbf{h})|^2 \le 1 \tag{3.15}$$

which is not very useful since it is an obvious consequence of the way that unitary structure factors are defined.

For a centrosymmetric structure

$$U(\mathbf{h}) = \sum_{j=1}^{N} n_j \cos(2\pi \mathbf{h} \cdot \mathbf{r}_j)$$

and the term in the summation is partitioned as $a_j = n_j^{1/2}$ and $b_j = n_j^{1/2} \cos(2\pi \mathbf{h} \cdot \mathbf{r}_j)$. As before

$$\sum_{j=1}^{N} |a_j|^2 = 1,$$

$$\sum_{j=1}^{N} |b_j|^2 = \sum_{j=1}^{N} n_j \cos^2(2\pi \mathbf{h} \cdot \mathbf{r}_j) = \tfrac{1}{2}\sum_{j=1}^{N} n_j [1 - \cos(4\pi \mathbf{h} \cdot \mathbf{r}_j)],$$

which leads to the inequality

$$U(\mathbf{h})^2 \le \tfrac{1}{2}[1 + U(2\mathbf{h})]. \tag{3.16}$$

If $|U(\mathbf{h})|$ and $|U(2\mathbf{h})|$ are sufficiently large then it is possible to show that $s(2\mathbf{h})$ must be $+1$. For example, if $|U(\mathbf{h})| = 0.5$ then the left-hand side of (3.16) equals 0.25. For $|U(2\mathbf{h})| = 0.6$ the right-hand side will be 0.8 for $s(2\mathbf{h}) = +1$ and 0.2 for $s(2\mathbf{h}) = -1$, showing that $s(2\mathbf{h})$ must be $+1$ to satisfy the inequality.

Another important type of inequality relationship may be found by taking the sum or difference of unitary structure factors for a centrosymmetric structure. From

$$U(\mathbf{h}) + U(\mathbf{h}') = \sum_{j=1}^{N} n_j \left[\cos(2\pi\mathbf{h}\cdot\mathbf{r}_j) + \cos(2\pi\mathbf{h}'\cdot\mathbf{r}_j) \right]$$

$$= \sum_{j=1}^{N} 2n_j \cos\left[\pi(\mathbf{h}+\mathbf{h}')\cdot\mathbf{r}_j \right] \cos\left[\pi(\mathbf{h}-\mathbf{h}')\cdot\mathbf{r}_j \right]$$

with the partitioning

$$a_j = (2n_j)^{\frac{1}{2}} \cos\left[\pi(\mathbf{h}+\mathbf{h}')\cdot\mathbf{r}_j \right],$$
$$b_j = (2n_j)^{\frac{1}{2}} \cos\left[\pi(\mathbf{h}-\mathbf{h}')\cdot\mathbf{r}_j \right]$$

there is obtained the inequality

$$[U(\mathbf{h}) + U(\mathbf{h}')]^2 \le [1 + U(\mathbf{h}+\mathbf{h}')][1 + U(\mathbf{h}-\mathbf{h}')]. \tag{3.17}$$

By an exactly similar process, starting with $U(\mathbf{h}) - U(\mathbf{h}')$ there is found

$$[U(\mathbf{h}) - U(\mathbf{h}')]^2 \le [1 - U(\mathbf{h}+\mathbf{h}')][1 - U(\mathbf{h}-\mathbf{h}')]. \tag{3.18}$$

If $U(\mathbf{h})$ and $U(\mathbf{h}')$ have the same sign, both positive or both negative, then (3.17) gives the greater value of the left-hand side and the $+$ signs on the right-hand side happen to be the sign of the product $U(\mathbf{h})U(\mathbf{h}')$. On the other hand, if $U(\mathbf{h})$ and $U(\mathbf{h}')$ are of opposite signs then (3.18) gives the greater left-hand-side value and the $-$ signs on the right-hand side are also the signs of $U(\mathbf{h})U(\mathbf{h}')$. From this consideration it is possible to combine (3.17) and (3.18) into a single inequality relationship

$$[|U(\mathbf{h})| + |U(\mathbf{h}')|]^2 \le [1 + s(\mathbf{h})s(\mathbf{h}')s(\mathbf{h}+\mathbf{h}')|U(\mathbf{h}+\mathbf{h}')|]$$
$$\times [1 + s(\mathbf{h})s(\mathbf{h}')s(\mathbf{h}-\mathbf{h}')|U(\mathbf{h}-\mathbf{h}')|]. \tag{3.19}$$

If the magnitudes of $U(\mathbf{h})$, $U(\mathbf{h}')$, $U(\mathbf{h}+\mathbf{h}')$ and $U(\mathbf{h}-\mathbf{h}')$ are sufficiently large then it can sometimes be shown that one or both of the products of three signs must be positive. For example with the magnitudes of all four U

terms equal to 0.5 (i.e. all the structure factors have one half of their maximum values) the value of the left-hand side of (3.19) is 1.0. Giving the products of signs all possible pairs of values, $+\,+$, $+\,-$, $-\,+$ and $-\,-$ it is found that only $+\,+$ satisfies the inequality so that both

$$s(\mathbf{h})s(\mathbf{h}')s(\mathbf{h}+\mathbf{h}') = +1$$

and

$$s(\mathbf{h})s(\mathbf{h}')s(\mathbf{h}-\mathbf{h}') = +1. \tag{3.20}$$

This confirms the statement made in §3.1.1 that a relationship of the form (3.3) may be true even if the corresponding structure factors do not have maximum possible values.

Combining (3.17) and (3.18) to give (3.19) is only valid if none of the U terms has zero magnitude, when a sign cannot be associated with it. If, in fact, one of the U terms on the left-hand side, say $U(\mathbf{h}')$, is zero then an interesting result may sometimes be found from considering (3.17) and (3.18) individually. For example consider

$$|U(\mathbf{h})| = 0.6, \ |U(\mathbf{h}')| = 0, \ |U(\mathbf{h}+\mathbf{h}')| = |U(\mathbf{h}-\mathbf{h}')| = 0.5$$

then the left-hand sides of both (3.17) and (3.18) will be 0.36. We now consider all possible combinations of signs for $U(\mathbf{h}+\mathbf{h}')$ and $U(\mathbf{h}-\mathbf{h}')$. If both signs are positive then substituting in equation (3.18) gives 0.25 on the right-hand side, which does not satisfy the inequality. Alternatively if both signs are negative then substituting in (3.17) will give 0.25 on the right-hand side which, again, does not satisfy the inequality. It has thus been shown that $U(\mathbf{h}+\mathbf{h}')$ and $U(\mathbf{h}-\mathbf{h}')$ cannot have the same sign. It can be checked that with opposite signs for the two U terms both inequalities give a right-hand side of 0.75, which means that they are both satisfied. It has thus been shown that, for this combination of unitary-structure-factor magnitudes,

$$s(\mathbf{h}+\mathbf{h}')s(\mathbf{h}-\mathbf{h}') = -1. \tag{3.21}$$

Harker–Kasper inequalities require large magnitudes of unitary structure factors for them to give useful information and hence they have little or nothing to contribute to the solution of structures of *current* interest. For a structure with 100 equal atoms in the unit cell we see from (3.13) that $|U|$ of 0.5 corresponds to $|E|$ of 5.0; it is unlikely that a structure will give one structure factor of this magnitude let alone several to incorporate in an inequality expression. For space groups of higher symmetry it is possible to generate inequality relationships that give information with somewhat smaller magnitudes. Some examples of these will be found in Woolfson (1961).

Shortly after the introduction of Harker–Kasper inequalities other forms of inequality relationships were found by other workers, e.g. Gillis (1948), Karle and Hauptman (1950) and Okaya and Nitta (1952a, b, c). The form of inequality given by Karle and Hauptman is quite different from all the others and we shall consider it in somewhat greater detail.

3.2.2 Determinant inequalities

In 1950 Karle and Hauptman used a theorem given by Toeplitz (1911) relating the Fourier coefficients of a real, non-negative periodic function to give inequality relationships, which we express in terms of unitary structure factors, of the form

$$
\begin{vmatrix}
1 & U(\mathbf{h}_1) & U(\mathbf{h}_2) & \cdots & U(\mathbf{h}_n) \\
U(\bar{\mathbf{h}}_1) & 1 & U(\mathbf{h}_2-\mathbf{h}_1) & & U(\mathbf{h}_n-\mathbf{h}_1) \\
U(\bar{\mathbf{h}}_2) & U(\mathbf{h}_1-\mathbf{h}_2) & 1 & & U(\mathbf{h}_n-\mathbf{h}_2) \\
\cdot & & & & \\
\cdot & & & & \\
\cdot & & & & \\
U(\bar{\mathbf{h}}_n) & U(\mathbf{h}_1-\mathbf{h}_n) & U(\mathbf{h}_2-\mathbf{h}_n) & & 1
\end{vmatrix} \geq 0.
$$

$$(3.22)$$

The whole determinant is defined by the leading row, the elements of which must all be different although they can be related by symmetry. The inequality is true, regardless of the order of the determinant, and we consider the case

$$
\begin{vmatrix}
1 & U(\mathbf{h}) & U(2\mathbf{h}) \\
U(\bar{\mathbf{h}}) & 1 & U(\mathbf{h}) \\
U(2\bar{\mathbf{h}}) & U(\mathbf{h}) & 1
\end{vmatrix} \geq 0. \qquad (3.23)
$$

Expanding the determinant one finds

$$1 - 2|U(\mathbf{h})|^2 - |U(2\mathbf{h})|^2 + U(\mathbf{h})^2 U(\overline{2\mathbf{h}}) + U(\bar{\mathbf{h}})^2 U(2\mathbf{h}) \geq 0. \qquad (3.24)$$

For a centrosymmetric structure, where all the structure factors are real, this gives

$$\left[1 - U(2\mathbf{h})\right]\left[1 + U(2\mathbf{h}) - 2U(\mathbf{h})^2\right] \geq 0$$

and since the first term must be non-negative this implies that

$$U(\mathbf{h})^2 \le \tfrac{1}{2}\left[1 + U(2\mathbf{h})\right]$$

which is the Harker–Kasper inequality (3.16). However, the determinants are also capable of giving quite new inequality relationships. Consider

$$\begin{vmatrix} 1 & U(\mathbf{h}) & U(\mathbf{h}') \\ U(\bar{\mathbf{h}}) & 1 & U(\mathbf{h}'-\mathbf{h}) \\ U(\bar{\mathbf{h}}') & U(\mathbf{h}-\mathbf{h}') & 1 \end{vmatrix} \ge 0. \qquad (3.25)$$

On expansion, for a centrosymmetric structure, this gives

$$1 - U(\mathbf{h})^2 - U(\mathbf{h}')^2 - U(\mathbf{h}-\mathbf{h}')^2 + 2U(\mathbf{h})U(\mathbf{h}')U(\mathbf{h}-\mathbf{h}') \ge 0$$

or

$$U(\mathbf{h})U(\mathbf{h}')U(\mathbf{h}-\mathbf{h}') \ge \tfrac{1}{2}\left[U(\mathbf{h})^2 + U(\mathbf{h}')^2 + U(\mathbf{h}-\mathbf{h}')^2 - 1\right]. \qquad (3.26)$$

This inequality has the advantage over (3.19) that it involves just three structure factors but on the other hand, in general, it is somewhat weaker. For example, with all four $|U|$ terms equal to 0.45 (3.19) shows that both relationships (3.20) must be true but the subset of three $|U|$ terms inserted in (3.26) does not determine with certainty the sign of $s(\mathbf{h})s(\mathbf{h}')s(\mathbf{h}-\mathbf{h}')$.

It is interesting to examine the expansion of (3.25) for a non-centrosymmetric structure. Bearing in mind that $U(\bar{\mathbf{h}}) = U(\mathbf{h})^*$ where * denotes 'complex conjugate' and that

$$U(\mathbf{h}) + U(\bar{\mathbf{h}}) = 2|U(\mathbf{h})| \cos\left[\phi(\mathbf{h})\right] \qquad (3.27)$$

we find

$$\cos\left[\phi(\mathbf{h}) - \phi(\mathbf{h}') - \phi(\mathbf{h}-\mathbf{h}')\right] \ge \frac{|U(\mathbf{h})|^2 + |U(\mathbf{h}')|^2 + |U(\mathbf{h}-\mathbf{h}')|^2 - 1}{2|U(\mathbf{h})U(\mathbf{h}')U(\mathbf{h}-\mathbf{h}')|}. \qquad (3.28)$$

With all $|U|$ terms equal to 0.6 this gives

$$\cos\left[\phi(\mathbf{h}) - \phi(\mathbf{h}') - \phi(\mathbf{h}-\mathbf{h}')\right] \ge 0.185$$

or

$$-79° \le \phi(\mathbf{h}) - \phi(\mathbf{h}') - \phi(\mathbf{h}-\mathbf{h}') \le 79°.$$

Here we see a constraint on the value of a combination of three phases for a non-centrosymmetric structure as previously, in (3.20), we have seen the derivation of specific information about a product of three signs of a centrosymmetric structure. It will be found that very often direct methods are concerned with the use of formulae which give estimates of combinations of phases or signs and we now consider the form that these combinations can take.

3.3 Structure invariants and seminvariants

3.3.1 Structure invariants

A crystal is a three-dimensional array of unit cells forming a periodic structure. The dimensions of a crystal (about 1 mm) are so large compared with the dimensions of a unit cell (1–10 nm) that it is reasonable to think of a crystal as an infinite periodic array. In fig. 3.2 there is shown a part of a two-dimensional periodic array; even if no unit-cell structure is shown the periodicity is obvious by inspection. In fig. 3.2 there are shown two choices of unit cell, each of which is a unit of the periodic structure. The point being made here is that even if the shape and size of the unit cell are specified, and they are the same as illustrated, the choice of origin is quite arbitrary and this means that the phases of structure factors are not invariant quantities completely defined by the total structure.

The structure-factor equation may be written in the form

$$F(\mathbf{h}) = |F(\mathbf{h})| \exp\left[i\phi(\mathbf{h})\right] = \sum_{j=1}^{N} f_j \exp(2\pi i \mathbf{h} \cdot \mathbf{r}_j) \qquad (3.29)$$

where the $\mathbf{r}_j$ correspond to a particular choice of origin. If it is now decided to change the origin to a point with vector position $\mathbf{R}$ with respect to the original origin then an atom that had vector position $\mathbf{r}_j$ with respect to the original origin will have vector position $\mathbf{r}_j - \mathbf{R}$ with respect to the new origin. The structure factor is now modified to

$$F'(\mathbf{h}) = |F'(\mathbf{h})| \exp\left[i\phi'(\mathbf{h})\right] = \sum_{j=1}^{N} f_j \exp\left[2\pi i \mathbf{h} \cdot (\mathbf{r}_j - \mathbf{R})\right]$$

$$= \exp(-2\pi i \mathbf{h} \cdot \mathbf{R}) \sum_{j=1}^{N} f_j \exp(2\pi i \mathbf{h} \cdot \mathbf{r}_j). \qquad (3.30)$$

A comparison of (3.29) and (3.30) shows that

$$|F'(\mathbf{h})| = |F(\mathbf{h})|,$$
$$\phi'(\mathbf{h}) = \phi(\mathbf{h}) - 2\pi\mathbf{h} \cdot \mathbf{R}. \qquad (3.31)$$

This shows that $|F(\mathbf{h})|$ is a *structure-invariant* quantity, i.e. independent of the choice of origin, whereas $\phi(\mathbf{h})$ is not.

We now examine the product of three structure factors

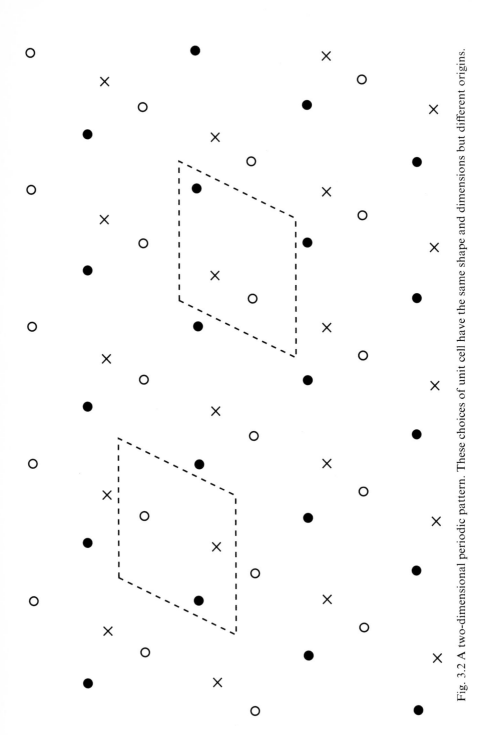

Fig. 3.2 A two-dimensional periodic pattern. These choices of unit cell have the same shape and dimensions but different origins.

$$F(\mathbf{h}_1)F(\mathbf{h}_2)F(\mathbf{h}_3) = |F(\mathbf{h}_1)F(\mathbf{h}_2)F(\mathbf{h}_3)| \exp\{i[\phi(\mathbf{h}_1) + \phi(\mathbf{h}_2) + \phi(\mathbf{h}_3)]\}.$$

From (3.31) it is readily found that if the origin is moved as previously described to give modified structure factors then

$$F'(\mathbf{h}_1)F'(\mathbf{h}_2)F'(\mathbf{h}_3) = F(\mathbf{h}_1)F(\mathbf{h}_2)F(\mathbf{h}_3)\exp\left[-2\pi i(\mathbf{h}_1 + \mathbf{h}_2 + \mathbf{h}_3)\cdot\mathbf{R}\right]. \quad (3.32)$$

The condition for the product of three structure factors to be a structure invariant is seen to be

$$\mathbf{h}_1 + \mathbf{h}_2 + \mathbf{h}_3 = 0. \tag{3.33}$$

It is this condition that has been present in the various conclusions drawn from inequality relationships, implicitly if not explicitly. The combination of three phases in (3.28) $\phi(\mathbf{h}) - \phi(\mathbf{h}') - \phi(\mathbf{h} - \mathbf{h}')$ can, from (1.24), be expressed as $\phi(\mathbf{h}) + \phi(-\mathbf{h}') + \phi(\mathbf{h}' - \mathbf{h})$ and the three indices have a null sum, as required by (3.33).

This principle can be extended to products of any number of structure factors in the form

$$\prod_{s=1}^{m} F(\mathbf{h}_s) \tag{3.34}$$

which will be a structure invariant under the condition

$$\sum_{s=1}^{m} \mathbf{h}_s = 0.$$

For $m = 1$ the structure factor, which is a structure invariant, is $F(0)$, which has a phase of zero for any choice of origin. For $m = 2$ the condition from (3.34) is that $\mathbf{h}_2 = -\mathbf{h}_1$ and the product of the two structure factors becomes $|F(\mathbf{h}_1)|^2$, which is phase-independent. Later we shall be referring to structure invariant quantities involving the product of more than three structure factors. It should be noted that all the relationships that have been derived for the usual structure factors, the F terms, will also apply to unitary or normalised structure factors.

The structure invariant conditions apply equally to centrosymmetric structures where it is more conventional to refer to the signs of structure factors. Because phases can only be 0 or π for centrosymmetric structures and because $0 \equiv -0$ and $\pi \equiv -\pi$, which means that $s(\mathbf{h}) = s(-\mathbf{h})$, the condition on indices can be somewhat obscured. The two products of signs given in (3.20) may be transformed as follows

$$s(\mathbf{h})s(\mathbf{h}')s(\mathbf{h} + \mathbf{h}') = s(\mathbf{h})s(\mathbf{h}')s(-\mathbf{h} - \mathbf{h}')$$

$$s(\mathbf{h})s(\mathbf{h}')s(\mathbf{h}-\mathbf{h}') = s(\mathbf{h})s(-\mathbf{h}')s(\mathbf{h}'-\mathbf{h}) \qquad (3.35)$$

and the sum of the three indices is null in each case.

The approach in direct methods, where structure invariant quantities, the $|E|$ terms, are used to estimate other structure invariant quantities such as

$$\Phi_3(\mathbf{h},\mathbf{h}') = \Phi(\mathbf{h}) - \phi(\mathbf{h}') - \phi(\mathbf{h}-\mathbf{h}') \qquad (3.36)$$

seems logically viable. However, we should distinguish quantities that are structure invariants from those that are *structure and enantiomorph invariants*. Enantiomorphs of a non-centrosymmetric structure are such that an atom with coordinate $\mathbf{r}_j$ in one structure will have coordinate $-\mathbf{r}_j$ in the other; corresponding structural units within enantiomorph structures display a mirror-image (left-hand–right-hand) relationship with respect to each other. Such pairs of structures have identical diffraction intensities and identical sets of $|E|$ terms but the phases of individual structure factors are the negatives of each other. This will also apply to the combination of phases in (3.35). Logically it is not possible to determine enantiomorph-dependent quantities just from the $|E|$ terms alone. What *can* be determined are quantities like $\cos\left[\Phi_3(\mathbf{h},\mathbf{h}')\right]$, which, since $\cos\theta = \cos(-\theta)$, are enantiomorph-independent. In fact, it can be seen in (3.28) that it is $\cos\left[\phi_3(\mathbf{h},\mathbf{h}')\right]$ that is involved in the inequality relationship and not $\phi_3(\mathbf{h},\mathbf{h}')$ itself.

3.3.2 *Structure seminvariants*

We saw illustrated in fig. 3.2 that, from the point of view of defining the periodic nature of a crystal structure, the choice of origin is arbitrary. However, except for the space group P1, the choice is always made to have some relationship to the symmetry elements present in the space group. If, for example, the structure has a centre of symmetry then it would be natural to choose it for an origin for then there would be the simplification that all the structure factors would be real. For another example we can take the space group $P2_1$ which has a twofold screw axis and, although the origin would not be uniquely defined, it would be natural to place it somewhere on the axis. The projection of the structure down the 2_1 axis, say along y, would then be centrosymmetric so that the $(h\ 0\ l)$ structure factors would be real.

In the space group $P\bar{1}$ there is a set of eight different centres of symmetry, as shown in fig. 3.3. The form of the structure factor equation

$$F(h\ k\ l) = \sum_{j=1}^{N} f_j \cos\left[2\pi(hx_j + ky_j + lz_j)\right] \qquad (3.37)$$

is the same for all the eight centres of symmetry as origin but the actual coordinates of the atoms will be different. If the origin is moved from point A to point B in fig. 3.3 then all the new x coordinates will be related to the old ones by

$$x_B = x_A - 0.5$$

while the y and z coordinates will be unchanged. Substituting in (3.37) it is found that

$$F_B(h\ k\ l) = (-1)^h F_A(h\ k\ l). \tag{3.38}$$

Similar considerations lead to the following table, which shows the sign of a positive structure factor for origin A for different parities of h, k and l when the origin is changed to other permitted positions. The symbols e and o refer to even and odd for h, k and l in that order.

Origin	Parity group							
	eee	oee	eoe	ooe	eeo	oeo	eoo	ooo
A	+	+	+	+	+	+	+	+
B	+	−	+	−	+	−	+	−
C	+	+	−	−	+	+	−	−
D	+	+	+	+	−	−	−	−
E	+	−	−	+	+	−	−	+
F	+	−	+	−	−	+	−	+
G	+	+	−	−	−	−	+	+
H	+	−	−	+	−	+	+	−

It will be seen that, in general, the sign depends on the choice of origin but it does not do so for the reflections with all even indices, i.e. those for which

$$(h\ k\ l)\,\text{modulo}\,(2\ 2\ 2) = (0\ 0\ 0).$$

Such reflections are said to be *structure seminvariants* since their phases (signs) are invariant for a *permitted* change of origin.

If for the spacegroup P$\bar{1}$ there is taken the product of three structure factors

$$F(h_1\ k_1\ l_1)F(h_2\ k_2\ l_2)F(h_3\ k_3\ l_3) \tag{3.39}$$

then if the origin is displaced to another centre of symmetry the sign of the product will be changed by

$$(\pm 1)^{h_1 + h_2 + h_3}(\pm 1)^{k_1 + k_2 + k_3}(\pm 1)^{l_1 + l_2 + l_3}$$

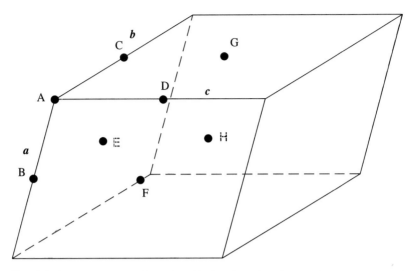

Fig. 3.3 A centrosymmetric unit cell showing the eight different centres of symmetry. E is at the centre of the *a–b* face and H is at the body centre.

where $+1$ or -1 depends on whether there was not or there was a shift of 0.5 in the appropriate direction. However, it can be seen that if

$$(h_1 + h_2 + h_3 \quad k_1 + k_2 + k_3 \quad l_1 + l_2 + l_3) \, \text{modulo} \, (2 \ 2 \ 2) = (0 \ 0 \ 0) \quad (3.40)$$

then the product of structure factors in (3.39) is a structure seminvariant. The difference between a structure invariant and a structure seminvariant is illustrated by the products of structure factors $F(1 \ 1 \ 4)F(0 \ 3 \ \bar{2})F(\bar{1} \ \bar{4} \ \bar{2})$ and $F(1 \ 1 \ 4)F(0 \ 3 \ \bar{2})F(3 \ 6 \ 2)$. The first product is a structure invariant because of the null sum of indices. It would be a structure invariant for both the space groups P1 and P$\bar{1}$ even if the origin were moved to a non-standard general position. The second product is a structure seminvariant because the sum of the indices is $(0 \ 0 \ 0) \, \text{modulo} \, (2 \ 2 \ 2)$ so that its phase (sign) will not be changed by moving to another standard origin. However, it would change if the origin were moved to a non-standard position.

Another example is the orthorhombic space group P$2_1 2_1 2_1$, which has three non-intersecting orthogonal sets of twofold screw axes. The arrangement of symmetry elements in relation to the chosen centre of symmetry is shown in fig. 1.24 as it appears in the *International Tables for Crystallography*. The eight permitted origins are situated halfway between the three pairs of screw axes and they are related by combinations of shifts of 0 or 0.5 along the axes, similarly to the centres of symmetry in P$\bar{1}$. In much the same way as for P$\bar{1}$ individual structure factors with indices such that $(h \ k \ l)$ modulo $(2\ 2\ 2) = (0\ 0\ 0)$ are structure seminvariants as would be a product of three structure factors as in (3.39) with indices satisfying relationship (3.40).

In a monograph written in 1953, Hauptman and Karle introduced the Σ_1 formula for estimating the values of individual phases that are structure seminvariants. Thus in the space group P$\bar{1}$ it appears in the form

$$s(2\mathbf{h}) \approx s[E(\mathbf{h})^2 - 1] \tag{3.41}$$

where $E(2\mathbf{h})$ has a large magnitude. The probability that $s(2\mathbf{h})$ is positive is given by

$$P_+(2\mathbf{h}) = \tfrac{1}{2} + \tfrac{1}{2}\tanh\left(\frac{\sigma_3}{\sigma_2^{3/2}} E(2\mathbf{h})\,[E(\mathbf{h})^2 - 1]\right) \tag{3.42}$$

where $\sigma_n = \sum_{i=1}^{N} z_i^n$ and z_i is the atomic number of the ith atom. For an equal atom structure $\sigma_3/\sigma_2^{3/2} = N^{-\frac{1}{2}}$, where N is the number of atoms in the unit cell.

This tends to be a weak relationship but a much stronger one is found in the presence of a 2_1 axis. Thus if there is a 2_1 axis along y then

$$s(2h\ 0\ 2l) \approx s\left(\sum_k (-1)^k [|E(h\ k\ l)|^2 - 1]\right) \tag{3.43}$$

with the probability that $s(2h\ 0\ 2l)$ is positive

$$P_+(2h\ 0\ 2l) = \tfrac{1}{2} + \tfrac{1}{2}\tanh\left(\frac{\sigma_3}{\sigma_2^{3/2}} |E(2h\ 0\ 2l)| \sum_k (-1)^k [|E(h\ k\ l)|^2 - 1]\right). \tag{3.44}$$

This can give quite good indications of signs for projection reflections and we shall give an example of its use in §3.5.1.

3.3.3 Selecting the origin and enantiomorph

The identification of structure invariants and seminvariants enables the origin and the enantiomorph of the structure to be fixed. We can illustrate this for space group P$\bar{1}$ by reference to the table in §3.3.2. If three structure factors are chosen from three different parity groups, excluding eee, such that

$$(h_1 + h_2 + h_3 \quad k_1 + k_2 + k_3 \quad l_1 + l_2 + l_3)\,\text{modulo}\,(2\ 2\ 2) \neq (0\ 0\ 0) \tag{3.45}$$

then the product of the three structure factors is not a structure seminvariant and can be either positive or negative. An arbitrary choice of the sign of each structure factor in the product can be made and for one of the eight possible origins the choice will be true. This is tantamount to fixing the origin at one of the eight possible positions. The condition can be put in a more concise way. If $h\,\mathrm{modulo}(2) = h$ then the condition for three suitable reflections to fix the origin is

$$
\begin{vmatrix}
h_1 & h_2 & h_3 \\
k_1 & k_2 & k_3 \\
l_1 & l_2 & l_3
\end{vmatrix} = \pm 1.
\tag{3.46}
$$

It is also possible to fix the origin for the space group P1. The condition that, when three reflections $(h_1\ k_1\ l_1)$, $(h_2\ k_2\ l_2)$ and $(h_3\ k_3\ l_3)$ are given arbitrary phases, they will uniquely fix the origin in the unit cell is

$$
\begin{vmatrix}
h_1 & h_2 & h_3 \\
k_1 & k_2 & k_3 \\
l_1 & l_2 & l_3
\end{vmatrix} = \pm 1.
\tag{3.47}
$$

For P1 there is also the possibility of fixing the enantiomorph. Since all phases are general and can be anywhere in the range 0 to 2π it is not possible to fix the enantiomorph by a specific allocation once the origin has been chosen. However, since a change of enantiomorph corresponds to a change of sign of all phases it is possible to fix the enantiomorph by specifying that some phase, which must not be 0 or π, is in the range 0 to π.

The space group $P2_12_12_1$ offers an interesting example of the way that the origin and enantiomorph can be fixed by the choice of four precise phases. With a standard origin, structure factors with indices $(h\ k\ 0)$, $(h\ 0\ l)$ and $(0\ k\ l)$ are either real, with phases 0 or π, or imaginary, with phases $\pm\pi/2$. The rule for this is that if the indices are treated in cyclic fashion then if the index following the zero is even the structure factor is real but if it is odd then the structure factor is imaginary. Three structure factors with indices satisfying the condition (3.46) can have their phases allocated arbitrarily to fix the origin. These may be chosen from the projection reflections, with one index zero – for example $F(5\ 0\ \bar{2})F(1\ 1\ 0)F(0\ 1\ 1)$, which may be given phases, 0, $\pi/2$ and $\pi/2$ to fix an origin at one of the eight standard positions. However, the selection of a fourth phase can also fix the enantiomorph. The trick here is to find a reflection that in product with two of those fixing the origin forms a structure seminvariant that is complex. For example, the indices in the product $F(1\ 1\ 0)F(0\ 1\ 1)F(3\ 0\ 5)$ do not satisfy (3.45) and

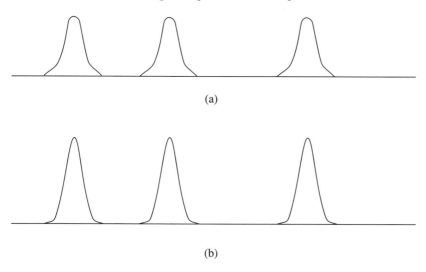

(a)

(b)

Fig. 3.4(a) Representation of electron density for a one-dimensional equal-atom structure. (b) The squared density.

so the product is a structure seminvariant. Since each of the individual structure factors is complex so is the product of three of them, which must therefore have value $\pi/2$ or $-\pi/2$. The two enantiomorphs will have opposite signs for the seminvariant so by selecting, say, $\pi/2$ the enantiomorph is being selected.

The theory of determining structure invariant and seminvariant products of structure factors and of origin and enantiomorph fixing to cover all space groups is extremely complex. A full treatment has been given by Hauptman and Karle (1953, 1956) with some corrections to their work by Lessinger and Wondratschek (1975).

3.4 Sign and phase relationships

3.4.1 Sayre's equation and sign relationships

A notable advance in direct methods was made by Sayre (1952) who showed that, under some circumstances, an exact equation linking structure factors can be found. In fig. 3.4(a) there is shown part of the electron density of a one-dimensional structure consisting of equal resolved atoms. The effect of squaring the electron density is shown in fig. 3.4(b). The squared density also shows equal resolved peaks and the structure factor for the squared density will be related to the original structure factors by

$$G(\mathbf{h}) = \frac{g(\mathbf{h})}{f(\mathbf{h})} F(\mathbf{h}) \tag{3.48}$$

where $f(\mathbf{h})$ and $g(\mathbf{h})$ are the scattering factors for the normal and squared electron density respectively.

We now make use of the well-known convolution theorem in Fourier transform theory, which states that

$$FT[F_1(\mathbf{r})F_2(\mathbf{r})] = FT[F_1(\mathbf{r})]^* FT[F_2(\mathbf{r})] \tag{3.49}$$

where $F_1(\mathbf{r})$ and $F_2(\mathbf{r})$ are two functions of position, $\mathbf{r}$, and * signifies convolution. The Fourier transform of the electron density is $F(\mathbf{h})$ so that, from (3.49), the Fourier transform of the electron density squared will be

$$G(\mathbf{h}) = F(\mathbf{h})^* F(\mathbf{h}). \tag{3.50}$$

Since $F(\mathbf{h})$ is a discrete function in reciprocal space, occurring only at points on a reciprocal lattice, with each structure factor associated with a volume of reciprocal space, $V^* = 1/V$, where V is the volume of the unit cell, then it is found that

$$G(\mathbf{h}) = \frac{1}{V} \sum_{\mathbf{k}} F(\mathbf{k})F(\mathbf{h} - \mathbf{k}). \tag{3.51}$$

Combining (3.48) and (3.51) leads to Sayre's equation

$$F(\mathbf{h}) = \frac{f(\mathbf{h})}{g(\mathbf{h})V} \sum_{\mathbf{k}} F(\mathbf{k})F(\mathbf{h} - \mathbf{k}). \tag{3.52}$$

In his original paper Sayre applied the equation to the centrosymmetric projection of a simple structure and found correct signs for the largest structure factors. However, the main outcome of this paper, and two other papers published alongside that of Sayre by Cochran (1952) and Zachariasen (1952), was a relationship between the signs of centrosymmetric structure factors

$$s(\mathbf{h}) \approx s(\mathbf{k})s(\mathbf{h} - \mathbf{k}) \tag{3.53}$$

where $\approx$ means 'probably equals'. This can be seen from (3.52); if $F(\mathbf{h})$ is a large structure factor then it is more likely than not that a large product on the right-hand side will have the same sign. Clearly the likelihood that the relationship is true will increase with the magnitudes of the three involved structure factors. Woolfson (1954) gave a formula for the probability that only applied to an equal-atom structure but Cochran and Woolfson (1955) gave a more general formula

$$P(\mathbf{h}, \mathbf{k}) = \tfrac{1}{2} + \tfrac{1}{2} \tanh \left(\frac{\sigma_3}{\sigma_2^{3/2}} |E(\mathbf{h})E(\mathbf{k})E(\mathbf{h}-\mathbf{k})| \right) \tag{3.54}$$

where σ_n was as defined in (3.42).

During the few years following the discovery of sign relationships they were used in various ways to solve centrosymmetric structures or centrosymmetric projections of non-centrosymmetric structures, e.g. for space groups $P2_12_12_1$ or $P2_1$; the limitations of data collection and computational facilities at that time meant that most structural work was done in projection.

As an illustration of a method, devised by Zachariasen, there is shown in fig. 3.5(a) one asymmetric unit of the $(0\ k\ l)$ section of the reciprocal lattice, with the unitary structure factors, for the structure of dicyclopentadienyldi-iron tetracarbonyl (Mills, 1958). The space group is $P2_1/a$ and the projection has the two-dimensional space group pgg for which the signs of structure factors are related as

$$s(0\ \bar{k}\ l) = (-1)^{k+l} s(0\ k\ l).$$

In fig. 3.5(b) the origin has been fixed by making two properly-chosen signs positive and five other structure factors of large $|U|$ have been allocated letter symbols, a b c d e, each representing either $+1$ or -1. For this simple structure Harker–Kasper inequality relationships could be used to find the certain signs of a number of other structure factors as shown in the figure. A transparent replica of fig. 3.5(b), placed over the figure with its origin placed over a position corresponding to a strong structure factor (index $\mathbf{h}$), will show overlapped all the pairs of letters or signs with indices $\mathbf{k}$ and $\mathbf{h}-\mathbf{k}$ that give a probable value of $s(\mathbf{h})$. Thus overlapping the point $(0\ 6\ 0)$ gives

$$s(0\ 6\ 0) \approx aa, ab, ab, ab, ab, ab, cd, e$$

and other overlaps give

$$s(0\ 8\ 0) \approx -, ab, ab, ab, ab, ab, ab, cd, cd,$$
$$s(0\ 10\ 0) \approx , -, +, ab, ab, ab, ab, cd.$$

The evidence suggests that $s(0\ 6\ 0) = s(0\ 8\ 0) = s(0\ 10\ 0) = ab$ and also that $ab = cd$ or $abcd = +1$. By a continuation of this process most of the

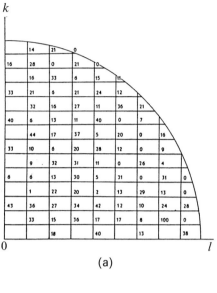

(a)

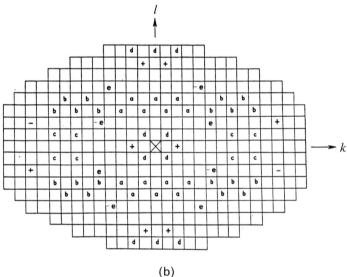

(b)

Fig. 3.5(a) One asymmetric unit of the (0 *k l*) reciprocal lattice section for dicyclopentadienyldi-iron tetracarbonyl, $Fe_2(CO)_4(C_5H_5)_2$ with the values of $100 \times |U(0\ k\ l)|$. (b) The position after fixing the origin, allocating five letter symbols to reflections with large $|U|$ and the application of inequality relationships.

large structure factors can have their signs indicated and in addition the sign symbols can either be found explicitly or relationships found between some of them. In the case of dicyclopentadienyldi-iron tetracarbonyl the reader may wish to confirm that all the sign symbols can be determined and that explicit signs can be found for all the large structure factors.

3.4.2 General phase relationships

The determinant-derived inequality (3.28) showed that a structure invariant combination of three general phases for a non-centrosymmetric structure could be constrained in value if the three corresponding normalised structure factors were all large. The larger the $|E|$ terms the tighter the constraint towards the value zero and, by analogy with the correspondence between inequality relationships and sign relationships for centrosymmetric structures we might expect that some probabilistic form of phase relationship should exist. Cochran (1955) showed that this was indeed so. The basic triplet phase relationship is of the form

$$\phi(\mathbf{h}_1) + \phi(\mathbf{h}_2) + \phi(\mathbf{h}_3) \approx 0 \quad \text{(modulo } 2\pi\text{)}, \tag{3.55}$$

where

$$\mathbf{h}_1 + \mathbf{h}_2 + \mathbf{h}_3 = 0$$

and $\approx$ indicates 'is distributed about'. Another form of the relationship is

$$\Phi_3(\mathbf{h}, \mathbf{h}') = \phi(\mathbf{h}) - \phi(\mathbf{h}') - \phi(\mathbf{h} - \mathbf{h}') \approx 0 \quad \text{(modulo } 2\pi\text{)}, \tag{3.56}$$

where $\Phi_3(\mathbf{h}, \mathbf{h}')$ is the structure invariant quantity given in (3.36).

For a given product of magnitudes of the three involved normalised structure factors the value of Φ_3 has a probability density given by

$$P(\phi_3) = \frac{1}{2\pi I_0[K(\mathbf{h}, \mathbf{k})]} \exp\left[K(\mathbf{h}, \mathbf{k}) \cos \Phi_3\right], \tag{3.57}$$

where $K(\mathbf{h}, \mathbf{k}) = 2(\sigma_3/\sigma_2^{3/2})|E(\mathbf{h})E(\mathbf{k})E(\mathbf{h} - \mathbf{k})|$, which is twice the argument in (3.54), and I_0 is the zeroth-order member of a family of modified Bessel functions, I_n. The form of $P(\Phi_3)$ is shown in fig. 3.6 for various values of K; it will be seen that for larger values of K the distribution is quite tightly constrained.

A rearrangement of (3.56) gives

$$\phi(\mathbf{h}) \approx \phi(\mathbf{k}) + \phi(\mathbf{h} - \mathbf{k}) \tag{3.58}$$

which shows that if the phases of two structure factors are known then the phase of the third one may be estimated. If the phases on the right-hand side

Direct methods

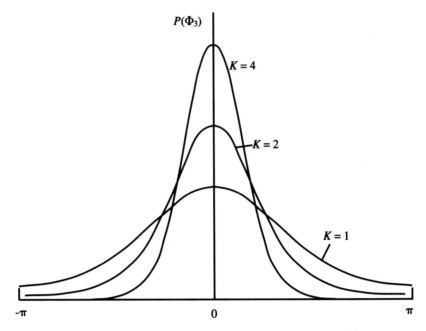

Fig. 3.6 The Cochran distribution for different values of K.

of (3.58) are known precisely then the variance in the estimate of $\phi(\mathbf{h})$ is found from a formula given by Karle and Karle (1966), which is

$$V = \frac{\pi^2}{3} + [I_0(K)]^{-1} \left(\sum_{n=1}^{\infty} \frac{I_{2n}(K)}{n^2} - 4 \sum_{n=0}^{\infty} \frac{I_{2n+1}(K)}{(2n+1)^2} \right) \qquad (3.59)$$

and is shown in graphical form in fig. 3.7(a). The corresponding curve for standard deviation, in degrees, is given in fig. 3.7(b).

During an application of direct methods there may be several pairs of 'known' phases, which individually give estimates of the phase $\phi(\mathbf{h})$ by the use of (3.58). These estimates will all differ and the question of how they should be combined to give an overall estimate was resolved by Karle and Hauptman (1956) when they introduced what is probably the most widely used formula in direct methods, the tangent formula

$$\tan\left[\phi(\mathbf{h})\right] = \frac{\sum\limits_{\mathbf{k}} K(\mathbf{h}, \mathbf{k}) \sin\left[\phi(\mathbf{k}) + \phi(\mathbf{h} - \mathbf{k})\right]}{\sum\limits_{\mathbf{k}} K(\mathbf{h}, \mathbf{k}) \cos\left[\phi(\mathbf{k}) + \phi(\mathbf{h} - \mathbf{k})\right]} = \frac{T(\mathbf{h})}{B(\mathbf{h})}. \qquad (3.60)$$

The form of the tangent formula given here differs a little from, but is equivalent to, that given by Karle and Hauptman. The reliability of the

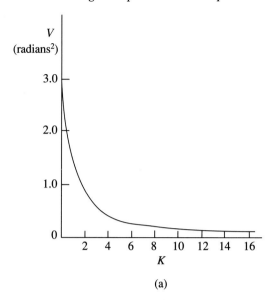

(a)

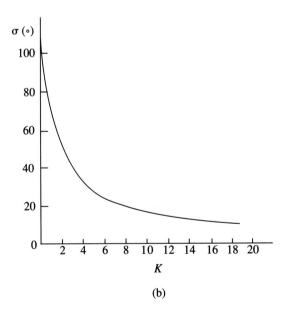

(b)

Fig. 3.7(a) The variance (in radians squared) of a single three-phase relationship as a function of K. (b) The standard deviation in degrees.

phase estimate given by (3.60) may be inferred from the diagrams given in fig. 3.8. In fig. 3.8(a) there are shown on Argand diagrams individual phase indications for $\phi(\mathbf{h}) - \phi(\mathbf{h})_1$, $\phi(\mathbf{h})_2$ and $\phi(\mathbf{h})_3$ – with moduli $K(\mathbf{h}, \mathbf{k}_1)$, $K(\mathbf{h}, \mathbf{k}_2)$ and $K(\mathbf{h}, \mathbf{k}_3)$, which indicate the variances of the estimates. In fig. 3.8(b) these are combined to give the tangent formula estimate $\phi(\mathbf{h})$; the variance of the overall estimate is related to $\alpha(\mathbf{h})$ in the same way as the individual variances are related to the values of K where

$$\alpha(\mathbf{h})^2 = T(\mathbf{h})^2 + B(\mathbf{h})^2. \tag{3.61}$$

If α replaces K in the variance curve, fig. 3.7(a), then the curve will indicate the variance of the estimate of $\phi(\mathbf{h})$ *on the assumption that all the phases on the right-hand side of the tangent formula are correct.* This assumption is rarely true so one must interpret estimated variances with great caution.

3.4.3 Other relationships between phases

We saw that the product of structure factors given in (3.34) is a structure invariant which, for $m = 4$, will lead to the following combination of phases, known as a quartet, being a structure invariant,

$$\Phi_4(\mathbf{h}, \mathbf{k}, \mathbf{l}) = \phi(\mathbf{h}) + \phi(\mathbf{k}) + \phi(\mathbf{l}) - \phi(\mathbf{h} + \mathbf{k} + \mathbf{l}). \tag{3.62}$$

If the four involved structure factors are all large then it has been shown by Hauptman (1975) that the probability density of $\phi_4(\mathbf{h}, \mathbf{k}, \mathbf{l})$ is given by

$$P[\phi_4(\mathbf{h}, \mathbf{k}, \mathbf{l})] = \frac{1}{2\pi I_0[B(\mathbf{h}, \mathbf{k}, \mathbf{l})]} \exp\{B(\mathbf{h}, \mathbf{k}, \mathbf{l}) \cos [\Phi_4(\mathbf{h}, \mathbf{k}, \mathbf{l})]\}, \tag{3.63}$$

where $B(\mathbf{h}, \mathbf{k}, \mathbf{l}) = \sigma_4/\sigma_2^2 |E(\mathbf{h})E(\mathbf{k})E(\mathbf{l})E(\mathbf{h} + \mathbf{k} + \mathbf{l})|$. The similarities with the Cochran formula, (3.57), will be evident and the form of the distribution and variance will be as in figs. 3.6 and 3.7 except that B plays the role of K.

If in addition to the four main $|E|$ terms being large there is also knowledge of some other E magnitudes then different estimates of the quartet value or its variance may be made. For example, suppose that it is known that $|E(\mathbf{h} + \mathbf{k})|$ is large. Then from (3.58)

$$\phi(\mathbf{h} + \mathbf{k}) \approx \phi(\mathbf{h}) + \phi(\mathbf{k}) \tag{3.64a}$$

$$\phi(\mathbf{h} + \mathbf{k}) \approx -\phi(\mathbf{l}) + \phi(\mathbf{h} + \mathbf{k} + \mathbf{l}). \tag{3.64b}$$

Subtracting (3.64b) from (3.64a) leads to

$$\Phi_4(\mathbf{h}, \mathbf{k}, \mathbf{l}) = \phi(\mathbf{h}) + \phi(\mathbf{k}) + \phi(\mathbf{l}) - \phi(\mathbf{h} + \mathbf{k} + \mathbf{l}) \approx 0 \text{ (modulo } 2\pi). \tag{3.65}$$

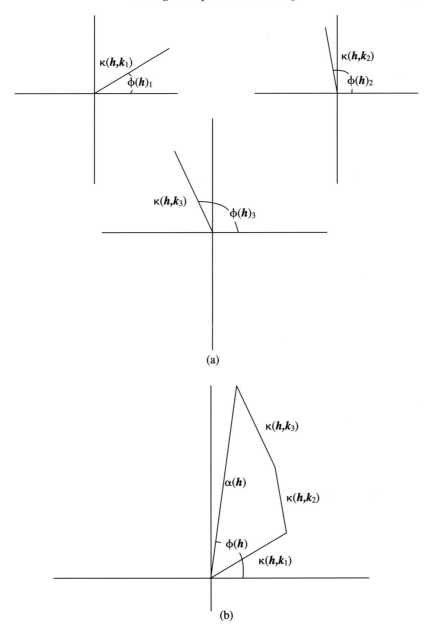

Fig. 3.8(a) Three phase indications represented on Argand diagrams with corresponding values of K. (b) The addition of the three contributions in the complex plane gives the tangent-formula indication $\phi(\mathbf{h})$.

This reinforces the conclusion that $\Phi_4(\mathbf{h}, \mathbf{k}, \mathbf{l})$ should be close to zero and, by symmetry, if $|E(\mathbf{k}+\mathbf{l})|$ and $|E(\mathbf{l}+\mathbf{h})|$ are also both large then the conclusion will be greatly reinforced.

The other extreme possibility is that $|E(\mathbf{h}+\mathbf{k})|$, $|E(\mathbf{k}+\mathbf{l})|$ and $|E(\mathbf{l}+\mathbf{h})|$, referred to as the cross-terms, are all very small. Schenk (1973a, b) and Schenk and de Jong (1973) put forward the idea, on rather heuristic grounds, that in such a situation the most probable value of the quartet would not be zero but π. Hauptman showed that this was true and gave a formula for the distribution of ϕ_4 conditional on the seven magnitudes

$$R_1 = |E(\mathbf{h})|, \ R_2 = |E(\mathbf{k})|, \ R_3 = |E(\mathbf{l})|, \ R_4 = |E(\mathbf{h}+\mathbf{k}+\mathbf{l})|,$$
$$R_{12} = |E(\mathbf{h}+\mathbf{k})|, \ R_{23} = |E(\mathbf{k}+\mathbf{l})|, \ R_{31} + |E(\mathbf{l}+\mathbf{h})|.$$

This is

$$P(\Phi_4 | R_1, R_2, R_3, R_4, R_{12}, R_{23}, R_{31}) =$$
$$\frac{1}{L} \exp(-2B' \cos \Phi_4) I_0(QR_{12}X_{12}) I_0(QR_{23}X_{23}) I(QR_{31}X_{31}) \quad (3.66)$$

where $Q = 2\sigma_3/\sigma_2^{3/2}$,

$$B' = \frac{1}{\sigma_2^3}(3\sigma_3^2 - \sigma_2\sigma_4)R_1R_2R_3R_4$$
$$X_{12} = [R_1^2R_2^2 + R_3^2R_4^2 + 2R_1R_2R_3R_4\cos(\Phi_4)]^{\frac{1}{2}}$$
$$X_{23} = [R_2^2R_3^2 + R_1^2R_4^2 + 2R_1R_2R_3R_4\cos(\Phi_4)]^{\frac{1}{2}}$$
$$X_{31} = [R_3^2R_1^2 + R_2^2R_4^2 + 2R_1R_2R_3R_4\cos(\Phi_4)]^{\frac{1}{2}}$$

and L is a normalising constant. Some four magnitude and seven magnitude distributions are shown in fig. 3.9. It will be seen that whereas the four-magnitude distribution is always monomodal with a peak at zero the seven magnitude distributions may also either be monomodal about π or bimodal. The quartets with most probable value π, called negative quartets because the cosine of the sum of four phases is negative, are not very common for most structures. However, they have sometimes been used effectively in the application of direct methods (see §3.5.8).

3.5 The application of direct methods

3.5.1 The symbolic-addition method

Although the Cochran relationship, in the form (3.55) or (3.58), was introduced in 1955 and the tangent formula (3.60) just one year later,

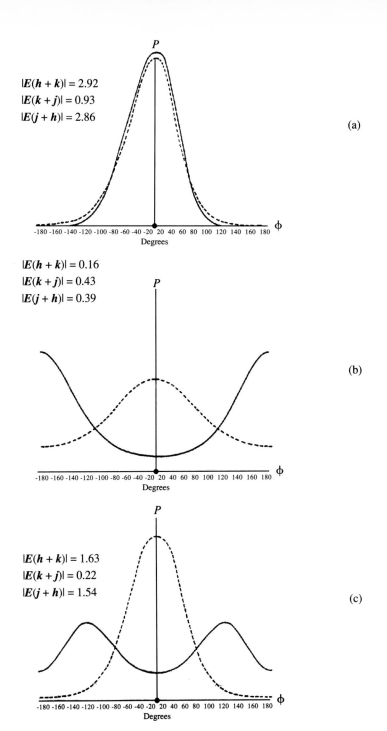

$|E(h + k)| = 2.92$
$|E(k + j)| = 0.93$
$|E(j + h)| = 2.86$

(a)

$|E(h + k)| = 0.16$
$|E(k + j)| = 0.43$
$|E(j + h)| = 0.39$

(b)

$|E(h + k)| = 1.63$
$|E(k + j)| = 0.22$
$|E(j + h)| = 1.54$

(c)

Fig. 3.9 The full lines represent the probability distribution of Φ_4 given only the values of the four principal normalised structure factors, all of which are large. The seven magnitude distributions are for: (a) three large cross terms, (b) three small cross terms, and (c) two large and one small cross term.

several years passed before the first systematic technique was designed to use these formulae to solve non-centrosymmetric crystal structures. In 1963 Karle and Karle introduced the symbolic-addition method, which they applied first to centrosymmetric structures and was similar to Zachariasen's use of symbols and also to a more general approach, which did not depend on inequality relationships, suggested by Woolfson (1961). However, the real breakthrough came when Karle and Karle (1964) showed that a method using symbols could be successful with a non-centrosymmetric structure, which they illustrated by the solution of the structure of L-arginine dihydrate in space group $P2_12_12_1$.

The first step in this process was to fix the origin by the allocation of phases to three projection reflections and symbols to four others, thus giving a starting set from which new phase information would be generated. Their starting set was

h	h	l	$|E|$	Phase or symbol
3	0	10	3.46	0
3	3	0	2.17	$-\pi/2$
3	0	1	2.77	$\pi/2$
2	12	0	3.21	p
2	10	0	2.31	s
4	0	14	2.56	m
3	8	3	2.31	a

p, s, m must be 0 or π

It was shown in §3.3.3 that it is possible in this space group to define the enantiomorph by selecting the phase of a suitable fourth axial reflection but Karle and Karle did not do this and fixed the enantiomorph later in the process. The first step in phase extension was the determination of the probable phase of the strong reflection (6 3 1) by

$$\phi(6\ 3\ 1)\approx\phi(3\ 0\ 1)+\phi(3\ 3\ 0)=\pi/2-\pi/2=0.$$

After two more steps the following consistent pair of indications was found

$$\phi(0\ 3\ 10)\approx\phi(\bar{3}\ 0\ \bar{1})+\phi(3\ 3\ 11)=-\pi/2+\pi=\pi/2,$$
$$\phi(0\ 3\ 10)\approx\phi(3\ 3\ 0)+\phi(\bar{3}\ 0\ 10)=-\pi/2+\pi=\pi/2.$$

The occurence of two or more consistent indications of a new phase is reassuring but after a few more steps the following pair of indications occurred:

$$\phi(1\ 0\ 9)\approx\phi(2\ 0\ 5)+\phi(\bar{1}\ 0\ 4)=(m-\pi/2)+(\pi+m)=\pi/2,$$
$$\phi(1\ 0\ 9)\approx\phi(3\ 0\ 1)+\phi(\bar{2}\ 0\ 8)=\pi/2+(\pi+m)=-\pi/2+m$$

followed by the four indications

$$\phi(4\ 3\ 7) \approx \phi(3\ 0\ \overline{10}) + \phi(1\ 3\ 17) = \pi + 0 = \pi,$$
$$\phi(4\ 3\ 7) \approx \phi(\overline{1}\ 0\ \ 4) + \phi(5\ 3\ \ 3) = (\pi - m) + m = \pi,$$
$$\phi(4\ 3\ 7) \approx \phi(1\ 0\ \ \overline{4}) + \phi(3\ 3\ 11) = (\pi + m) + \pi = m,$$
$$\phi(4\ 3\ 7) \approx \phi(\overline{2}\ 0\ \ 8) + \phi(6\ 3\ \ \overline{1}) = (\pi + m) + \pi = m.$$

There is a strong indication that $m = \pi$ and later other indications are found for symbols. At one stage it is strongly indicated that the symbol a is either 0 or π and later it is found that $\phi(0\ 8\ 7) = \pi + a$. Taking $a = 0$ gives $\phi(0\ 8\ 7) = \pi$, which fixes the enantiomorph since reflection $(0\ 8\ 7)$ combines with two origin-fixing reflections to give a structure seminvariant, the value of which is $\pm \pi/2$.

At this stage a number of seminvariant projection reflections had phases determined in terms of symbols and Karle and Karle used the Σ_1 relationship (3.43) to determine probable signs for these reflections and hence for the symbols. Moderately strong Σ_1 indications were found for the following reflections:

h	k	l	Symbol	P_+	Probable sign
2	12	0	p	0.14	−
2	10	0	s	0.15	−
4	8	0	$\pi + p$	0.72	+
0	12	8	p	0.27	−
0	10	14	s	0.28	−

Although none of these individual indications is very strong the aggregate indications that p and s are both negative (phase indications π) are quite reliable.

Because the only general reflection, $(3\ 8\ 3)$, had its phase symbol determined as 0 all the 137 phase indications from the symbolic addition process gave phases of 0, $\pi/2$, π or $3\pi/2$. At this stage the tangent formula was used both to refine phases to self-consistency and also to estimate new phases for all reflections with $|E| > 1.0$. Eventually 400 phases were so determined and used to produce the map shown in fig. 3.10; this map, an E-map, is calculated with Fourier coefficients of magnitude $|E|$, which tends to give sharper and better resolved peaks than usual electron-density maps.

This first determination of a non-centrosymmetric structure by direct methods was a milestone in their development and led to a stream of new ideas in the next few years.

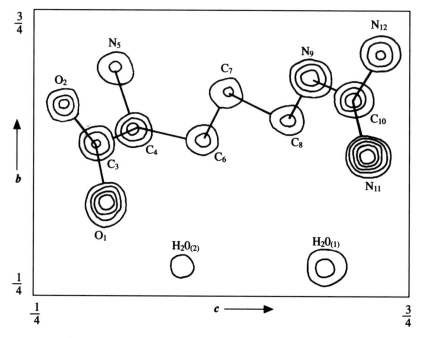

Fig. 3.10 The *E*-map for L-arginine dihydrate with phases found by symbolic addition for 400 reflections with $|E| > 1.0$.

3.5.2 *Other ways of using symbols*

In 1955 Cochran and Douglas designed the first programme to apply direct methods to centrosymmetric structures on the simple computers available at that time. Their process involved the use of symbols and in 1968 Germain and Woolfson applied similar concepts in producing a very effective method for centrosymmetric structures called LSAM (Logical Symbolic Addition Method), which also built on the work of Karle and Karle. LSAM was designed to be quite automatic and not to require the intervention of the user. The data input is the basic minimum required to solve the problem and the output a set of signs with an associated figure of merit. In the latest version the final output is a representation of the structure output on line-printer paper (see §3.5.3). The stages in LSAM are as follows.

3.5.2.1 *Data processing*

Raw data (intensities or structure factors) are processed to give normalised structure factors. This is usually done using the Wilson plot. The mean observed intensity as a function of Bragg angle θ is given by

$$\langle I\rangle_{\text{obs}}^{\theta}=k\Sigma\exp\left(-2B\frac{\sin^2\theta}{\lambda^2}\right),\tag{3.67}$$

where k is a scaling constant, Σ is defined in (2.31a) and B, the temperature factor, is presumed to be the same for all atoms and isotropic. From (3.67) we find

$$\log\left(\frac{\langle I\rangle_{\text{obs}}^{\theta}}{\Sigma}\right)=\log k-2B\frac{\sin^2\theta}{\lambda^2}.\tag{3.68}$$

The slope and intercept of the straight line Wilson plot of $\log(\langle I\rangle_{\text{obs}}^{\theta}/\Sigma)$ against $\sin^2\theta/\lambda^2$ give the values of $-2B$ and $\log k$. To find the normalised structure factor corresponding to $I(\mathbf{h})_{\text{obs}}$ one can use

$$|E(\mathbf{h})|^2=\frac{I(\mathbf{h})_{\text{obs}}}{\langle I\rangle_{\text{obs}}^{\theta}},\tag{3.69}$$

where the divisor is calculated from the values of B, k and the value of Σ for the appropriate θ. As found in this way the average values of $|E|^2$ plotted against $\sin^2\theta/\lambda^2$ show bumps characteristic of prominent interatomic vectors in the structure. A simpler method of deriving values of $|E|$, sometimes called the K-curve method, involves using the actual observed value of $\langle I\rangle_{\text{obs}}^{\theta}$ as the divisor in (3.70). This gives $\langle|E|\rangle^2=1$ over the whole range of scattering angles and some workers prefer this method on theoretical grounds.

3.5.2.2 Σ_2 listing

A subset of the E terms of largest magnitude is chosen for carrying out sign determination, the number being dependent on the number of independent non-hydrogen atoms in the asymmetric unit. This number is in the range 7–10 per independent atom, the factor being smaller for larger structures. There are then found the sets of three reflections whose indices are linked as in (3.20) and which form a sign relationship such that one sign may be estimated if the other two are known, as in (3.53). The complete list of the sets of three reflections forms what is called a Σ_2 list, which notation derives from the 1953 Hauptman and Karle monograph.

3.5.2.3 Origin and symbol allocation

A weight is allocated to each reflection according to the extent to which it participates in sign relationships with other reflections. Reflections with high weight are used to fix the origin and other high-weight reflections have their signs indicated by letter symbols – A, B, C etc. Normally between three and six symbols are used depending on the size of the structure.

3.5.2.4 Symbolic addition

There now begins a process of symbolic addition, which uses

$$s(\mathbf{h}) = s\left(\sum_{\mathbf{k}} E(\mathbf{k}) E(\mathbf{h} - \mathbf{k}) \right) \tag{3.70}$$

where the probability for the equal-atom case that the sign given by the summation is correct is

$$P(\mathbf{h}) = \tfrac{1}{2} + \tfrac{1}{2} \tanh \left(N^{-\frac{1}{2}} |E(\mathbf{h}) \sum_{\mathbf{k}} E(\mathbf{k}) E(\mathbf{h} - \mathbf{k})| \right). \tag{3.71}$$

One can only combine together symbolic indications that are the same and if, say, two indications are found – $A\,B$ with probability 0.98 and $-\,C$ with probability 0.92 then the indication with higher probability is accepted.

In this process of finding new sign indications only the indication of the highest probability is taken at each step. When this highest probability falls below some preset limit, which might be 0.975 for a small structure but 0.90 for a large structure, then this stage is complete.

3.5.2.5 Finding relationships between symbols

The sign indications are now inserted in all the relationships in the Σ_2 list and this leads to relationships between the symbols. For example, in a structure, Benperidol, space group $R\bar{3}$, $a = 36.62$ Å, $b = 7.705$ Å, $Z = 18$, solved by LSAM three structure factors, indicated with code numbers 1, 292 and 295 gave a sign relationship

$$s(1)s(292)s(295) \approx +1$$

or

$$A \times AC \times ABCD \approx +1$$

or

$$ABD \approx +1. \tag{3.72}$$

Several other relationships gave the same indication. If their probabilities were $P_1, P_2 \ldots P_n$ then the overall probability that $ABD \approx +1$ can be shown to be

$$P_{\text{comb}} = \frac{\prod\limits_{j=1}^{n} P_j}{\prod\limits_{j=1}^{n} P_j + \prod\limits_{j=1}^{n}(1 - P_j)}. \tag{3.73}$$

For the Benperidol case the resultant list of relationships was

$$CD \approx +1 \ (s_1) \qquad\qquad BCD \approx +1 \ (s_9)$$
$$BC \approx +1 \ (s_2) \qquad\qquad C \approx +1 \ (s_{10})$$
$$ABD \approx +1 \ (s_3) \qquad\qquad ABCD \approx +1 \ (s_{11})$$
$$ACD \approx +1 \ (s_4) \qquad\qquad AC \approx +1 \ (s_{11})$$
$$ABC \approx +1 \ (s_5) \qquad\qquad AD \approx +1 \ (s_{12})$$
$$A \approx +1 \ (s_6) \qquad\qquad AB \approx +1 \ (s_{13})$$
$$BD \approx +1 \ (s_7) \qquad\qquad D \approx +1 \ (s_{14})$$
$$B \approx +1 \ (s_8)$$

where the relationships all have probabilities very close to unity.

3.5.2.6 Developing sets of signs

LSAM next looks for a linearly independent set of these relationships such that inversion of the equations can give the letter symbols in terms of the s terms. The four equations that it found were numbers 1, 2, 3 and 8 and the inversion of the equations gives

$$A = s_1 s_2 s_3,$$
$$B = s_8,$$
$$C = s_2 s_8,$$
$$D = s_1 s_2 s_8. \tag{3.74}$$

All sets of signs, five in all, are found for the letter symbols such that not more than one of the four s terms in (3.74) is negative. Then LSAM goes back to the origin-fixing and starting set of structure factors and, with each possible set of signs for the symbols, develops new signs, one at a time from (3.70) accepting only that with the highest probability. Each set of signs is developed until the probability for a new sign falls below some limit, which was 0.80 for Benperidol.

3.5.2.7 Figures of merit

To test the finally developed sets of signs for plausibility three figures of merit are used, each of which is high if the sign relationships are working well. These are

$$M_1 = \sum_h \sum_k s(\mathbf{h})s(\mathbf{k})s(\mathbf{h}-\mathbf{k}), \tag{3.75}$$

$$M_2 = \sum_h \sum_k |E(\mathbf{h})E(\mathbf{k})E(\mathbf{h}-\mathbf{k})|s(\mathbf{h})s(\mathbf{k})s(\mathbf{h}-\mathbf{k}), \tag{3.76}$$

$$M_3 = \sum_h \sum_k P(\mathbf{h},\mathbf{k})s(\mathbf{h})s(\mathbf{k})s(\mathbf{h}-\mathbf{k}), \tag{3.77}$$

where $P(\mathbf{h},\mathbf{k})$ is given in (3.54). The figures of merit for each of the five starting points are given below with the total number of signs determined (for 300 structure factors in the system).

Set	A	B	C	D	M_1	M_2	M_3	Number determined
1	+	+	+	+	1643	1750	1449	295
2	+	−	−	−	1408	1355	1087	285
3	−	+	+	+	1011	812	612	258
4	−	+	−	−	1195	913	704	293
5	−	+	+	−	1220	933	720	294

The first set corresponds to all signs being positive, which would give a large peak at the origin and is unacceptable for this structure. Of the remaining sets the second has better figures of merit than the others and does, in fact, give the solution.

3.5.2.8 Showing the structure

With the signs from the second set an E-map is calculated and the highest peaks, in number somewhat greater than the number of independent atoms, are found. These are interpreted in terms of reasonable chemical molecules or fragments and are presented in a favourable projection on the line-printer of the computer, together with tables giving putative bond lengths and bond angles. It remains to the user to connect up the numbers on the paper to produce a representation of the structure. This final stage is shown for Benperidol in fig. 3.11.

LSAM is an extremely effective programme for centrosymmetric structures. Its basic strength is that it systematically allows for some failure of sign relationships and is thus a multi-solution method. It has been described

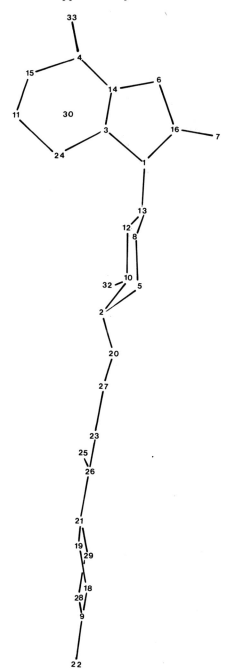

Fig. 3.11 The line-printer output from LSAM for the structure of Benperidol.

in great detail because it is a prototype for many other successful automatic computer-based methods of solving crystal structures.

Another successful and complete computer package, which uses symbols for solving non-centrosymmetric structures, is SIMPEL, developed by H. Schenk and his co-workers (Schenk and Kiers, 1984). It contains features similar to those in symbolic addition and in LSAM but also uses quartet relationships, both positive and negative, quintet relationships involving five phases and Σ_1 relationships. It is a single-solution procedure, which has a fairly high success rate with moderate-size structures and takes much less computer time than multi-solution procedures.

3.5.3 *MULTAN*

In the symbolic addition method it is not possible to combine indications in terms of different sets of symbols for a particular phase or sign. For example, if two separate indications for a phase are $a+b$ and $c+d$ then the new phase indication cannot simply be written as $\frac{1}{2}(a+b+c+d)$. If $a=b=\pi/2$ and $c=d=-\pi/2$ then the two individual indications are π and $-\pi$, which are equivalent, whereas the false combination would give 0. This is the problem of the '2π ambiguity', which prevents symbols representing phases being manipulated like algebraic symbols.

For this reason it is advantageous to introduce phases rather than symbols as early as possible into the phase-determining process as separate indications can then always be combined with the tangent formula (3.60). This led to the concept of MULTAN (MULtisolution TANgent-formula method) by Germain and Woolfson (1968). Just as with symbolic addition there is set up a starting set of reflections, some of which can be given origin- and enantiomorph-defining phase values, others may have phases reliably indicated by, for example, the Σ_1 formula while additional reflections are included to provide an adequate basis for phase extension. It is in relation to this last category of reflections that MULTAN solutions differ from symbolic addition for their phases are given explicit values rather than being represented by symbols. A special reflection, according to its type, may have one or other of a pair of values such as $(0, \pi)$ or $(\pm\pi/2)$ while a general reflection is given one or other of the quadrant values ($\pm\pi/4$ or $\pm 3\pi/4$). Fixing values in quadrants ensures that one of the phase values has an error less than or equal to 45° with an average phase error of $22\frac{1}{2}°$. The MULTAN process starts with all possible combinations of phase for the unknowns and extends and refines phases with the tangent formula.

Another important feature of MULTAN is that it uses a weighted tangent formula, rather than (3.60), of the form

$$\tan\left[\phi(\mathbf{h})\right] = \frac{\sum_{\mathbf{k}} w(\mathbf{k})w(\mathbf{h}-\mathbf{k})K(\mathbf{h},\mathbf{k})\sin\left[\phi(\mathbf{k})+\phi(\mathbf{h}-\mathbf{k})\right]}{\sum_{\mathbf{k}} w(\mathbf{k})w(\mathbf{h}-\mathbf{k})K(\mathbf{h},\mathbf{k})\cos\left[\phi(\mathbf{k})+\phi(\mathbf{h}-\mathbf{k})\right]} = \frac{T(\mathbf{h})}{B(\mathbf{h})}, \quad (3.78)$$

where

$$w(\mathbf{h}) = \text{minimum of }\left[0.2\alpha(\mathbf{h}),\ 1.0\right], \quad (3.79)$$

and $\alpha(\mathbf{h})$ is defined in (3.61).

The general structure of MULTAN is similar to that of LSAM, which has already been described, and is given below with details only for those features that differ from the LSAM approach.

3.5.3.1 Data processing

This is carried out as for LSAM.

3.5.3.2 Selecting large reflections for phasing

MULTAN selects a number of reflections whose phases are to be determined. An empirical set of rules is used to determine this number which, for a triclinic crystal, is

$$4 \times \text{number of independent atoms} + 100.$$

In general, the higher the symmetry the greater is the number of reflections for a given number of independent atoms. The reflections picked are those with the largest $|E|$ values and they are allocated code numbers, 1, 2, 3 etc. by which they are subsequently identified in the programme.

At this stage there are also selected 50–100 reflections of small (ideally zero) magnitude for use later in a figure of merit.

3.5.3.3 Σ_2 listing

The sets of three reflections whose phases are linked by the three-phase relationships (3.55) are found. In addition there are found pairs of reflections that contribute to the right-hand sides of Sayre's equation (3.52) for equations where the structure factor on the left-hand side corresponds to one of the small reflections selected in the previous stage.

3.5.3.4 Finding the starting set

This is done via a process called convergence (Germain, Main and Woolfson, 1970). At each stage of the process that reflection is identified which is least strongly linked to the remaining reflections via phase relationships and it, and all relationships involving it, are usually eliminated from the system. However, if without this reflection enantiomorph and origin definition would not be possible then it becomes a member of the starting set. Again, if it is completely unattached to the remaining reflections, so that its phase could not be estimated even if all the phases in the residual set were known, then again it becomes a member of the starting set. At the end of the convergence procedure the starting set contains reflections that define the origin and enantiomorph and other reflections, which, if their phases were known, would tend to lead to a strong development of phase information for the whole set of reflections.

3.5.3.5 Allocation of initial phases

In earlier versions of MULTAN this was done by permutation of quadrant phases or other specific values, as previously described. Thus for L-arginine dihydrate, space group $P2_12_12_1$, the starting phases would be permutations of

h	k	l	$\lvert E\rvert$	Phase
3	0	10	3.46	0
3	3	0	2.17	$-\pi/2$
3	0	1	2.77	$\pi/2$
2	12	0	3.21	$0, \pi$
2	10	0	2.31	$0, \pi$
4	0	14	2.56	$0, \pi$
3	8	3	2.31	$\dfrac{\pi}{4}, \dfrac{3\pi}{4}, \dfrac{5\pi}{4}, \dfrac{7\pi}{4}$

giving $32 = 2 \times 2 \times 2 \times 4$ starting sets. With a usual run of MULTAN an extra reflection would have been added to the first three so as to define both the origin and the enantiomorph. For example, $\lvert E(0\ 3\ 1)\rvert = 1.86$ and choosing its phase as $\pi/2$ would fix the enantiomorph (§3.3.3).

Later phase permutation was replaced by a process depending on the concept of *magic integers* first introduced by White and Woolfson (1975). They showed that a set of integers $m_1, m_2 \ldots m_n$ could be found such that n phases, in cycles, could be approximately represented by a single variable x in the range 0 to 1 via

$$\phi_i \cong m_i x \quad \text{(modulo 1)} \quad i = 1 - n. \tag{3.80}$$

As an example we could have $\{m\} = \{5, 7, 8, 9\}$ and values of x would be taken as $1/64$ to $63/64$ in steps of $1/32$, i.e. 32 sets of values for the phases. With $\phi_1 = 0.13$, $\phi_2 = 0.74$, $\phi_3 = 0.99$ and $\phi_4 = 0.59$ and with $x = 25/64$ the values of $m_i x$ are 0.953, 0.734, 0.125 and 0.516 giving errors, in cycles, of

0.177, 0.006, 0.135 and 0.074 respectively. This converts to a mean phase error of 35°.

By the use of magic integers the number of trials for a given number of starting-set reflections to be phased can be much reduced or, alternatively, for a given number of trials the size of the starting set can be substantially increased. This has the penalty of a somewhat larger expected mean phase error for the best starting-set phases but, as it turns out, the MULTAN process is easily able to accommodate this.

3.5.3.6 Tangent-formula refinement

MULTAN normally uses the weighted tangent formula (3.78). The weighting scheme reduces the influence of poorly determined phases and so tends to give a faster and more reliable phase development. However, there is a tendency for phases to go to values that make the phase relationships work too well – for P$\bar{1}$, for example, all phases being zero makes all relationships equal to zero and all tangent formulae perfectly consistent. This will happen for all *symmorphic space groups*, i.e. those without a translational symmetry element. In this case it is better to use a weighting scheme devised by Hull and Irwin (1978). This attempts to restrict the value of $\alpha(\mathbf{h})$, as defined in (3.61), to its expected value $\alpha(\mathbf{h})_{est}$, which may be found from

$$\alpha(\mathbf{h})_{est}^2 = \sum_k K(\mathbf{h},\mathbf{k})^2 + \sum_k \sum_l K(\mathbf{h},\mathbf{k})K(\mathbf{h},\mathbf{l})\eta(\mathbf{h},\mathbf{k})\eta(\mathbf{h},\mathbf{l}), \qquad (3.81)$$
$$\scriptstyle k \neq l$$

where

$$\eta(\mathbf{h},\mathbf{k}) = \frac{I_1[K(\mathbf{h},\mathbf{k})]}{I_0[K(\mathbf{h},\mathbf{k})]}, \qquad (3.82)$$

which is a ratio of modified Bessel functions.

The form of the Hull–Irwin weighting scheme is

$$w(\mathbf{h}) = A\exp(-x^2)\int_0^x \exp(-t^2)\,\mathrm{d}t, \qquad (3.83)$$

where $x = \alpha(\mathbf{h})/\alpha(\mathbf{h})_{est}$ and A is adjusted to make the maximum of $w(\mathbf{h})$ equal to unity. The form of $w(\mathbf{h})$ as a function of x is shown in fig. 3.12.

This weighting scheme is particularly valuable in preventing the loss of the enantiomorph in a symmorphic space group. Unweighted tangent formula refinement of phases in P2_1, for example, can sometimes introduce a mirror plane through the molecule perpendicular to the 2_1 axis. With the

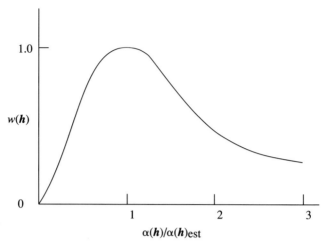

Fig. 3.12 The Hull–Irwin weighting function.

Hull–Irwin scheme this is avoided without the need to take any special precautions.

3.5.3.7 Figures of merit

The figures of merit (FOMs) described for LSAM are not applicable to non-centrosymmetric structures and an important part of the development of MULTAN was finding effective FOMs. Those used are

$$\text{ABSFOM} = \frac{\sum_{\mathbf{h}} \left[\alpha(\mathbf{h}) - \alpha(\mathbf{h})_r\right]}{\sum_{\mathbf{h}} \left[\alpha(\mathbf{h})_{est} - \alpha(\mathbf{h})_r\right]}, \tag{3.84}$$

where $\alpha(\mathbf{h})_{est}$ is defined in (3.81) and $\alpha(\mathbf{h})_r$, the expected value of $\alpha(\mathbf{h})$ with random phases, is given by

$$\alpha(\mathbf{h})_r^2 = \sum_{\mathbf{k}} K(\mathbf{h}, \mathbf{k})^2. \tag{3.85}$$

ABSFOM will tend to be unity for phases giving the expected values of $\alpha(\mathbf{h})$ and zero for random phases. In practice, because the tangent formula tends to give phases that make phase relationships hold too well, a good set of phases may give ABSFOM in the range 1.0–1.4.

The second FOM, ψ_0, owes much to the so-called zero check used by Cochran and Douglas (1955). This is

$$\psi_0 = \frac{\sum_{\mathbf{h}} \left| \sum_{\mathbf{k}} E(\mathbf{k})E(\mathbf{h}-\mathbf{k}) \right|}{\sum_{\mathbf{h}} \left(\sum_{\mathbf{k}} \left| E(\mathbf{k})E(\mathbf{h}-\mathbf{k}) \right|^2 \right)^{1/2}}, \qquad (3.86)$$

where the summation terms include the large structure factors whose phases are being determined but the outer summations are over values of $\mathbf{h}$ for which $E(\mathbf{h})$ is small, ideally zero. The ψ_0 figure of merit is expected to be small for a correct set of phases and also for a random set of phases. However, once an initially random phase set has been processed by the tangent formula it is no longer random and ψ_0 is quite effective in picking out plausible sets of phases.

The final FOM is

$$R_\alpha = \frac{\sum_{\mathbf{h}} \left| \alpha(\mathbf{h}) - \alpha(\mathbf{h})_{\text{est}} \right|}{\sum_{\mathbf{h}} \alpha(\mathbf{h})_{\text{est}}}, \qquad (3.87)$$

which is a residual between the actual and estimated values of α.

With the three figures of merit available for a number of different phase sets it is possible to rank them in order of plausibility by a combined figure of merit CFOM defined as

$$\text{CFOM} = w_1 \frac{\text{ABSFOM} - \text{ABSFOM}_{\text{min}}}{\text{ABSFOM}_{\text{max}} - \text{ABSFOM}_{\text{min}}} + w_2 \frac{(\psi_0)_{\text{max}} - \psi_0}{(\psi_0)_{\text{max}} - (\psi_0)_{\text{min}}}$$
$$+ w_3 \frac{(R_\alpha)_{\text{max}} - R_\alpha}{(R_\alpha)_{\text{max}} - (R_\alpha)_{\text{min}}}. \qquad (3.88)$$

The subscripts max and min refer to the value for the phase set with the greatest or least FOM and experience has shown that good values for the three weights are 0.6, 1.2 and 1.2 respectively. A set of phases with the best value for each of the FOMs would have CFOM $= 3.0$; with the worst value for each then CFOM $= 0$.

3.5.3.8 Calculating and interpreting an E-map

For the set of phases with the highest value of CFOM, or any other chosen by the user, an *E*-map is calculated and a number of the highest peaks are automatically found; this number is usually 25% more than the number of independent non-hydrogen atoms in the structure. These peaks are separated into potentially bonded clusters and then stereochemical criteria of acceptable bond lengths and angles are used to identify possible molecular fragments. In the usual version of MULTAN the positions of atomic peaks are presented on line-printer paper in a favourable projection together with tables of interpeak distances and bond angles, assuming that peaks are atomic positions, and also suggested molecular fragments.

Often the whole structure, or the greater part of it, is revealed and then the complete structure can be found and refined by conventional methods. If a small fragment is found then there are effective ways of developing it to the complete structure, as are described in chapter 7. A number of fragment-extending processes are contained in the MULTAN package.

The availability of MULTAN and other similar systems, notably SHELX (Sheldrick, 1975), had an important influence on the development of structural crystallography because it put into the hands of the crystallographer versatile tools, which could solve most small molecule structures quite automatically. From the time these programmes became available the actual solution of crystal structures became a much less demanding part of the crystallographer's task.

3.5.4 *Other strategies*

The first application of the magic-integer concept was not to improve the efficiency of a MULTAN starting set, as has already been described, but completely to replace the initial steps of MULTAN by an alternative procedure (White and Woolfson, 1975). For a particular magic integer sequence a number of different variables can be used so that, for example, a sequence of five magic integers and three variables

$$\begin{pmatrix} 16 \\ 24 \\ 28 \\ 30 \\ 31 \end{pmatrix} (x \ y \ z) \qquad (3.89)$$

may be used to represent 15 phases and some values of x, y and z should give reasonable approximations to those phases. With phases represented by magic integers it is now possible to represent three-phase relationships by magic integers. From (3.89) we may find $\phi_1 \approx 16x$, $\phi_7 \approx 24y$ and $\phi_{10} \approx 31y$ so that a relationship expressed in cycles in a way that it might appear in a Σ_2 list,

$$\phi_1 - \phi_7 + \phi_{10} + 0.5 \approx 0 \quad \text{(modulo 1)}$$

appears in magic-integer form as

$$16x + 7y + 0.5 \approx 0 \quad \text{(modulo 1)}. \tag{3.90}$$

In general a three-phase invariant appears in the form

$$Hx + Ky + Lz + b \approx 0 \quad \text{(modulo 1)}$$

and if there are M relationships linking the phases of the 15 reflections plus others that fix the origin and enantiomorph or are known from Σ_1 relationships, then the function

$$\psi(x, y, z) = \sum_{r=1}^{M} K_r \cos \left[2\pi (H_r x + K_r y + L_r z + b_r) \right] \tag{3.91}$$

should be a maximum for values of (x, y, z) that give good approximations to the phases.

The function $\psi(x, y, z)$ is an easily computed Fourier summation and the peaks in this function give values of x, y and z that can be used to generate trial values for the phases as starting points for phase extension using the tangent formula.

Declercq, Germain and Woolfson (1975) extended the magic-integer approach to give a very effective procedure, MAGIC. They extended the representation of phases by magic integers by using a process called the P–S (primary–secondary) method. A number of phases, the primaries, are directly represented by magic integers as in (3.89) or are known from origin and enantiomorph definition or Σ_1 relationships. Strong three-phase relationships are then found, which link two primaries and a third reflection (a secondary); the secondary can now be expressed in magic-integer form. The ψ map (3.91) may then be calculated with all relationships linking the primaries and secondaries.

MAGIC has been shown to be able to solve structures for which MULTAN had failed. Further improvements by Hull, Viterbo, Woolfson and Zhang (1981) and by Zhang and Woolfson (1982) gave an even more powerful procedure, MAGEX.

The reason for the success of MAGIC and MAGEX was that they enabled a larger number of reflections and relationships to be used right at the beginning of the phase determining process. In addition the phase-relationships were employed all together in a ψ-map rather than one or a few at a time in a chain process as in MULTAN. For this reason a few unreliable relationships would not affect the process so badly since it depended on the statistical behaviour of a large number of relationships.

Conversely, the presence of one or two unreliable relationships in the early stages of MULTAN could have a catastrophic effect.

Another separate line of development in direct methods was initiated by Woolfson (1977) who was investigating an alternative to the tangent formula as a procedure for phase refinement. Once phase estimates are available for a set of reflections it was suggested that refinement could proceed by the least-squares solution of a set of linear equations of the form

$$K(\phi_p \pm \phi_q \pm \phi_r + b) = Kn \tag{3.92}$$

where the quantity in parentheses is a phase relationship as it appears in a MULTAN Σ_2 list, with phases expressed in cycles, K is the normal weight associated with the relationship and n is some integer. With phases available, estimates of the integers can be made and the complete set of equations of the form

$$\mathbf{A}\Phi = \mathbf{c}, \tag{3.93}$$

where Φ and $\mathbf{c}$ are vectors and $\mathbf{A}$ is a matrix, gives a least-squares solution

$$\Phi = (\mathbf{A}^{\mathrm{T}}\mathbf{A})^{-1}\mathbf{A}^{\mathrm{T}}\mathbf{c}. \tag{3.94}$$

With new phase estimates new estimates could be made for the right-hand-side integers in the relationships and a cyclic process of refinement is available. Tests showed that such a procedure is always better than the tangent formula in terms of the final mean phase error and it is sometimes very much better.

If the initial phases are of poor quality then it is desirable to use additional weights other than just the values of K. For example, if with the current phases the estimate of the right-hand side was 0.5 then clearly the nearest integer is indeterminate and the appropriate relationship should have zero weight. If the value was 0.48 then, while 0 is the nearest integer there is some uncertainty and a low weight would seem appropriate. The problem here is that the weights, and therefore the matrix $\mathbf{A}$, would vary from one cycle to the next, requiring a time-consuming different matrix inversion, $(\mathbf{A}^{\mathrm{T}}\mathbf{A})^{-1}$, each time. To solve this problem a pseudo-weighting scheme was introduced which made each equation of the form

$$K(\phi_p \pm \phi_q \pm \phi_r + b) = K(n + 4\alpha^3), \tag{3.95}$$

where α is the deviation of the current estimate of the relationship value from the nearest integer and so is in the range $-\frac{1}{2} \le \alpha < \frac{1}{2}$. For the $\alpha = 0.5$ the value in parentheses on the right-hand side of (3.95) is $(n + 0.5)$; the equation is effectively neutralized since it has no tendency to move the values of ϕ from their current values. For $\alpha = 0.48$, the right-hand side is $(n + 0.442)$, which means that the equation will have little effect. This scheme proved to be very successful and the whole refinement process could be completed with only a single matrix inversion.

A natural property to investigate for any refinement process is its radius of convergence and this was done for the linear equations (Baggio, Woolfson, Declercq and Germain, 1978). Random errors were imposed on the correct phases, with increasing r.m.s. values and the ability of the refinement process to finish with a substantially correct phase set, taken as being one with a mean phase error less than $30°$, was explored. Several trial structures were used, including 3-chloro-1,3,4-triphenyl-azetidin-2-one, a structure that was not straightforward to solve with MULTAN. Increasing the r.m.s. error gave the surprising result that even with random phases some fraction of trial starting points were capable of successful refinement to substantially correct phases. This provided the basis of a new method, YZARC, which started with a basis of about 100 reflections to which random values were given followed by linear-equations refinement. Phase extension was then carried out with the tangent formula. Like MAGIC and MAGEX, YZARC treated simultaneously a large number of reflections and relationships and so was not prone to the problems of the effect of individual unreliable relationships in a chain process.

The random-start idea was taken to its ultimate limit by Yao (1981) in the RANTAN procedure in which random phases were allocated to *all* reflections in the system, which were then refined by the tangent formula. The success of the method depends on the use of a controlled weighting scheme during the refinement, where the weights are incorporated in the tangent formula as shown in (3.78) and calculated as in (3.79). Weights are allocated to the initial phases as follows:

origin fixing phases	weight = 1.00
enantiomorph-fixing phase with special value	weight = 0.99
enantiomorph-fixing phase with general value	weight = 0.85
random phase	weight = 0.25.

For a Σ_1-determined phase, where the estimated value had a probability P, the weight used was $2P - 1$. In the subsequent tangent-formula refinement a phase was not changed until the new phase estimate had a weight calculated by (3.79) that was larger than its initial weight; thereafter it was allowed to change normally.

RANTAN is usually the default system in using the MULTAN programme. It is very straightforward and it is not necessary to go through the convergence procedure since phases are developed all together and not by a chain process. In addition the RANTAN philosophy of starting with random phases allocated to all the reflections for which phases are required tends to be that used with programmes that have followed MULTAN.

An important MULTAN-style programme, which has facilities for solving structures with pseudo-symmetry is called SAPI; it is fully described in §7.3 and §7.4.1.

3.5.5 SAYTAN

The basic idea behind the method SAYTAN is that a good set of phases should satisfy a system of Sayre equations

$$E(\mathbf{h}) = \frac{K}{g(\mathbf{h})} \sum_{\mathbf{k}} E(\mathbf{k}) E(\mathbf{h} - \mathbf{k}), \tag{3.96}$$

where $g(\mathbf{h})$ is the scattering factor for squared-density atoms and K is an overall scaling constant, which allows for the fact that only structure factors with large magnitude are included on the right-hand side. In deriving a phasing equation Debaerdemaeker, Tate and Woolfson (1985, 1988) started with the following residual for a system of Sayre equations:

$$R = \sum_{\mathbf{h}} \left| g(\mathbf{h}) E(\mathbf{h}) - KG(\mathbf{h}) \right|^2 ; \quad G(\mathbf{h}) = \sum_{\mathbf{k}} E(\mathbf{k}) E(\mathbf{h} - \mathbf{k}). \tag{3.97}$$

As a condition that R should be a minimum it is necessary that

$$\frac{\partial R}{\partial \phi(\mathbf{h})} = 0 \text{ for all } \mathbf{h}$$

and this leads to the Sayre-equation tangent formula

$$\phi(\mathbf{h}) = \text{phase of} \left(\sum_{\mathbf{l}} [g(\mathbf{h}) + g(\mathbf{l}) + g(\mathbf{h} - \mathbf{l})] E(\mathbf{l}) E(\mathbf{h} - \mathbf{l}) \right.$$
$$\left. - 2K \sum_{\mathbf{l}} E(\mathbf{h} - \mathbf{l}) \sum_{\mathbf{k}} E(\mathbf{k}) E(\mathbf{l} - \mathbf{k}) \right). \tag{3.98}$$

A distinctive feature of the Sayre-equation tangent formula is that it can use the information from Sayre equations (3.96) for which the values of $|E(\mathbf{h})|$ are small, ideally zero. These are the terms used in the ψ_0 figure of merit (3.86) but here they are used actively to determine phases rather than passively to test their plausibility.

The 'zero' terms can be incorporated into the total residual R with their own weight and this leads to a modified form of (3.98), namely

$$\phi(\mathbf{h}) = \text{phase of} \left(\sum_{\mathbf{l}} [g(\mathbf{h}) + g(\mathbf{l}) + g(\mathbf{h} - \mathbf{l})] E(\mathbf{l}) E(\mathbf{h} - \mathbf{l}) \right.$$
$$\left. - 2K[q_B(\mathbf{h}) + w_Z q_W(\mathbf{h})] \right), \qquad (3.99)$$

where

$$q_B(\mathbf{h}) = \sum_{\mathbf{l}} E(\mathbf{h} - \mathbf{l}) \sum_{\mathbf{k}} E(\mathbf{k}) E(\mathbf{l} - \mathbf{k}) \qquad (3.100a)$$

the indices $\mathbf{h}$, $\mathbf{k}$ and $\mathbf{l}$ all referring to strong reflections that are to be phased, and

$$q_Z(\mathbf{h}) = \sum_{\mathbf{l}_Z} E(\mathbf{h} - \mathbf{l}_Z) \sum_{\mathbf{k}} E(\mathbf{k}) E(\mathbf{l}_Z - \mathbf{k}) \qquad (3.100b)$$

where the indices $\mathbf{l}_Z$ are for weak reflections that are not to be phased. Usually $q_B(\mathbf{h})$ and $q_Z(\mathbf{h})$ are referred to as the 'big' quartets and 'zero' quartets respectively.

The triplet term in (3.99) may also be divided into two parts

$$t_1(\mathbf{h}) = g(\mathbf{h}) \sum_{\mathbf{l}} E(\mathbf{l}) E(\mathbf{h} - \mathbf{l}),$$
$$t_2(\mathbf{h}) = [g(\mathbf{l}) + g(\mathbf{h} - \mathbf{l})] \sum_{\mathbf{l}} E(\mathbf{l}) E(\mathbf{h} - \mathbf{l}). \qquad (3.101)$$

We now find that

$$t_2(\mathbf{h}) - 2Kq_B(\mathbf{h}) = \sum_{\mathbf{l}} \{ E(\mathbf{h} - \mathbf{l})[g(\mathbf{l})E(\mathbf{l}) - KG(\mathbf{l})] + E(\mathbf{l})[g(\mathbf{h} - \mathbf{l})E(\mathbf{h} - \mathbf{l})$$
$$- KG(\mathbf{h} - \mathbf{l})] \}. \qquad (3.102)$$

Since the two terms on the right-hand side both contain Sayre equations as factors then for a set of phases satisfying Sayre's equation

$$t_2(\mathbf{h}) - 2Kq_B(\mathbf{h}) = 0. \qquad (3.103)$$

This suggests the following form for SAYTAN:

$$\phi(\mathbf{h}) = \text{phase of} \{ t_1(\mathbf{h}) + w_B[t_2(\mathbf{h}) - 2Kq_B(\mathbf{h}) - 2Kw_Z q_Z(\mathbf{h})] \} \qquad (3.104)$$

where w_B and w_Z are weights under the control of the user. The default value of w_B is zero and this is found to give good results in practice while at the same time it makes unnecessary the time-consuming computation of $q_B(\mathbf{h})$.

An important point to be considered is the selection of the scaling factor K in (3.97). A value that seems reasonable is that which gives a minimum of R, which is found from

$$\frac{\partial R}{\partial K} = -2T + 2KQ = 0, \qquad (3.105)$$

where T is the real part of $\sum_h g_h(\mathbf{h})E(\mathbf{h})^*G(\mathbf{h})$ and

$$Q = \sum_h |G(\mathbf{h})|^2 = \sum_h \sum_l \sum_{l'} |E(\mathbf{l})E(\mathbf{h}-\mathbf{l})E(\mathbf{l}')E(\mathbf{h}-\mathbf{l}')| \cos[\phi(\mathbf{l}) + \phi(\mathbf{h}-\mathbf{l})$$
$$- \phi(\mathbf{l}') - \phi(\mathbf{h}-\mathbf{l}')].$$

This gives

$$K = T/Q \qquad (3.106)$$

but since T and Q are functions of the phases then in the early stages of using the Sayre-equation tangent formula, when phases are random or nearly so, the values of T and Q will both be small and random. A good strategy has been found, namely to use

$$K = T/\langle Q \rangle$$

where $\langle Q \rangle$ is an estimated value of Q expected for a good set of phases. Early in the phase development process T is small and so the process is dominated by the three-phase relationships as in the application of the usual tangent formula (3.60). However, after a few cycles K becomes appreciable and then the extra information of the quartet terms is deployed.

The final point concerns the selection of the weighting factor w_Z in (3.103). An empirical value, which works well over a large range of conditions, is

$$w_Z = \frac{5}{1 + 8r} \qquad (3.107)$$

where r is the ratio of the number of contributions to ψ_0 terms to the number of three-phase relationships. Sometimes the effectiveness of SAYTAN, particularly for large structures, depends critically on the value of w_Z and it can be advantageous to try a range of values if it does not succeed first time.

Since it uses extra information SAYTAN is more effective than MULTAN, either giving a solution in fewer trials or giving a solution where MULTAN would not. In particular it has been shown to be capable of giving an *ab initio* solution of a small protein containing over 300 independent non-hydrogen atoms (Woolfson and Yao, 1990; Mukherjee and Woolfson, 1993). Attempts to use it on larger proteins, using up to 5000 trials, have not been successful and it seems likely that for larger proteins some new approach is required.

3.5.6 The SIR approach

A number of different direct methods packages have been produced from time to time, mostly strongly related to each other. A package that contains unique features is the SIR programme produced by Giacovazzo and his collaborators (Burla, Camalli, Cascarano, Giacovazzo, Polidori, Spagna and Viterbo, 1989).

In §3.4.3 there was described the theory underlying the negative quartet. With only four magnitudes, $|E(\mathbf{h})|$, $|E(\mathbf{k})|$, $|E(\mathbf{l})|$ and $|E(\mathbf{h}+\mathbf{k}+\mathbf{l})|$, taken into account the four-phase structure invariant

$$\Phi_4(\mathbf{h},\mathbf{k},\mathbf{l}) = \phi(\mathbf{h}) + \phi(\mathbf{k}) + \phi(\mathbf{l}) - \phi(\mathbf{h}+\mathbf{k}+\mathbf{l})$$

has a monomodal probability distribution centred on zero. However, bringing in the magnitudes of the four cross-terms $|E(\mathbf{h}+\mathbf{k})|$, $|E(\mathbf{k}+\mathbf{l})|$ and $|E(\mathbf{l}+\mathbf{h})|$ can modify the probability distribution in various ways – causing it to peak more sharply about zero or causing it to peak about π or making it bimodal and symmetrical with peaks at $\pm\phi_p$, where ϕ_p lies between 0 and π. In general we may write the probability density for a structure invariant Φ in the form of a conditional distribution function $P(\Phi|\{R\})$, where $\{R\}$ will usually be a set of normalized structure-factor magnitudes. This set of magnitudes will inevitably include the magnitudes associated with the phases appearing in Φ but may include extra ones as well – for example, the cross-term magnitudes for the quartet. Both Hauptman (1976) and Giacovazzo (1977, 1980) have considered the selection of effective sets of extra reflections. Hauptman has employed what he calls the *neighbourhood principle* by which he finds successive sets of reflections, called *neighbourhoods*, each contained within the following one and from the magnitudes of which it is possible to estimate the cosine of a structure invariant or seminvariant. Giacovazzo's formulation is similar but different; he uses a general theory, the *theory of representations*, which also gives a sequence of subsets of reflections, each contained in the succeeding one, such that their magnitudes lead to statistically more significant estimates of Φ as the subsets become larger. Giacovazzo's extensive work in this field has been incorporated in a successful computer package called SIR and while, inevitably, it may take longer to run than some other packages, because of the extra theory it calls upon, it can sometimes succeed where another method would fail.

Direct methods

We can follow the general pattern of representation theory by considering the first and second representations for a three-phase invariant in a particular case for the space group P3$_1$. For this space group equivalent positions are (x, y, z), $(-y, x-y, z+\frac{1}{3})$ and $(y-x, -x, z+\frac{2}{3})$. The symmetry operations in going from (x, y, z) to the final two points are shown by

$$\begin{pmatrix} -y \\ x-y \\ z+\frac{1}{3} \end{pmatrix} = \begin{pmatrix} 0 & -1 & 0 \\ 1 & -1 & 0 \\ 0 & 0 & 1 \end{pmatrix} \begin{pmatrix} x \\ y \\ z \end{pmatrix} + \begin{pmatrix} 0 \\ 0 \\ \frac{1}{3} \end{pmatrix} = R_2 \mathbf{r} + t_2, \tag{3.108}$$

$$\begin{pmatrix} y-x \\ -x \\ z+\frac{2}{3} \end{pmatrix} = \begin{pmatrix} -1 & 1 & 0 \\ -1 & 0 & 0 \\ 0 & 0 & 1 \end{pmatrix} \begin{pmatrix} x \\ y \\ z \end{pmatrix} + \begin{pmatrix} 0 \\ 0 \\ \frac{2}{3} \end{pmatrix} = R_3 \mathbf{r} + t_3, \tag{3.109}$$

where the R terms are the rotational parts and the t terms the translational parts of the symmetry operations.

We now consider the three-phase invariants

$$\Phi_3 = \phi(3\ 0\ \bar{3}) + \phi(\bar{3}\ 3\ 1) + \phi(0\ \bar{3}\ 2) \tag{3.110}$$

and also

$$\Phi_3' = \phi(3\ 0\ \bar{3}) + \phi(0\ \bar{3}\ 1) + \phi(\bar{3}\ 3\ 2). \tag{3.111}$$

Since

$$(0\ \bar{3}\ 1) = (\bar{3}\ 3\ 1)R_3$$

$(0\ \bar{3}\ 1)$ is an equivalent reflection to $(\bar{3}\ 3\ 1)$ and the phases are related by

$$\phi(0\ \bar{3}\ 1) = \phi(\bar{3}\ 3\ 1) - 2\pi(\bar{3}\ 3\ 1)t_3$$
$$= \phi(\bar{3}\ 3\ 1) - 4\pi/3. \tag{3.112}$$

Similarly, since $(\bar{3}\ 3\ 2) = (0\ \bar{3}\ 2)R_2$

$$\phi(\bar{3}\ 3\ 2) = \phi(0\ \bar{3}\ 2) - 2\pi(0\ \bar{3}\ 2)t_2$$
$$= \phi(0\ \bar{3}\ 2) - 4\pi/3. \tag{3.113}$$

Inserting (3.112) and (3.113) into (3.111) and comparing with (3.110) gives

$$\Phi_3' = \Phi_3 - 2\pi/3. \tag{3.114}$$

Clearly this does not allow both the relationships (3.110) and (3.111) to obey the Cochran relationship (3.55) and be close to zero. Giacovazzo has shown that where there are several symmetry related three phase invariants with $-\Delta_j$ as the phase shift of the jth one compared with the basic triplet the expected value of the basic triplet, Δ, is given by

$$\tan \Delta = \left(\sum_j \sin(\Delta_j) \right) \bigg/ \left(\sum_j \cos(\Delta_j) \right). \tag{3.115}$$

The probability distribution of the value of the basic invariant now becomes

$$P(\Phi_3) = \frac{1}{2\pi I_0[K'(\mathbf{h}, \mathbf{k})]} \exp\left[K'(\mathbf{h}, \mathbf{k}) \cos(\Phi_3 - \Delta) \right]. \tag{3.116}$$

where $K'(\mathbf{h}, \mathbf{k})$ is a slightly modified form of $K(\mathbf{h}, \mathbf{k})$. The information from (3.114) gives $\pi/3$ as the expectation value of Φ_3 and this value can be incorporated into the tangent formula when used in the SIR procedure.

The second representation of Φ_3 is

$$\Phi_3'' = \Phi_3' + \phi(\mathbf{k}R_i) - \phi(\mathbf{k}R_i) \qquad i = 1, 2, \ldots m \tag{3.117}$$

where Φ_3' is any of the triplets of the first representation and $\mathbf{k}$ is a free vector. The right-hand side of (3.117) is a special kind of quintet relationship (a structure invariant involving five phases) for which there are $6m$ cross-term magnitudes involved in its conditional probability. These are

$$|E(\mathbf{h}_1 + \mathbf{k}R_1)||E(\mathbf{h}_1 + \mathbf{k}R_2)| \ldots |E(\mathbf{h}_1 + \mathbf{k}R_m)|,$$
$$|E(\mathbf{h}_1 - \mathbf{k}R_1)||E(\mathbf{h}_1 - \mathbf{k}R_2)| \ldots |E(\mathbf{h}_1 - \mathbf{k}R_m)|,$$
$$|E(\mathbf{h}_2 + \mathbf{k}R_1)||E(\mathbf{h}_2 + \mathbf{k}R_2)| \ldots |E(\mathbf{h}_2 + \mathbf{k}R_m)|,$$
$$|E(\mathbf{h}_2 - \mathbf{k}R_1)||E(\mathbf{h}_2 - \mathbf{k}R_2)| \ldots |E(\mathbf{h}_2 - \mathbf{k}R_m)|,$$
$$|E(\mathbf{h}_3 + \mathbf{k}R_1)||E(\mathbf{h}_3 + \mathbf{k}R_2)| \ldots |E(\mathbf{h}_3 + \mathbf{k}R_m)|,$$
$$|E(\mathbf{h}_3 - \mathbf{k}R_1)||E(\mathbf{h}_3 - \mathbf{k}R_2)| \ldots |E(\mathbf{h}_3 - \mathbf{k}R_m)|.$$

For each free vector $\mathbf{k}$ a conditional probability for the quintet can be found and since the quintet must have the same value as Φ_3' this means that many probability distributions can be found for Φ_3'. These probability distributions may be considered as independent and their product, suitably scaled, gives an overall probability distribution for Φ_3', which might have quite a small variance.

The SIR programme makes extensive use of conditional distributions for both invariants and seminvariants found by representation theory and in that sense it differs from the more usual MULTAN approach.

3.5.7 SHELXS-86 and phase annealing

The direct-methods programme SHELXS-86 has many features in common with MULTAN but does have some quite distinctive features. In particular it incorporates negative quartets (§3.4.3) directly into the phasing procedure. It does this by using a modified tangent formula

$$\phi(\mathbf{h})_{\text{new}} = \text{phase of } [\alpha(\mathbf{h}) - \eta(\mathbf{h})], \qquad (3.118)$$

where

$$\alpha(\mathbf{h}) = \sum_{\mathbf{k}} K(\mathbf{h}, \mathbf{k}) \exp[\phi(\mathbf{k}) + \phi(\mathbf{h} - \mathbf{k})],$$

$$\eta(\mathbf{h}) = \sum_{\mathbf{k}} \sum_{\mathbf{l}} g(\mathbf{h}, \mathbf{k}, \mathbf{l}) B(\mathbf{h}, \mathbf{k}, \mathbf{l}) \exp[\phi(\mathbf{k}) + \phi(\mathbf{l}) + \phi(\mathbf{h} - \mathbf{k} - \mathbf{l})].$$

The negative quartets included are just those for which the three cross-terms are all small and $B(\mathbf{h}, \mathbf{k}, \mathbf{l})$ is defined for equation (3.63). The positive constants $g(\mathbf{h}, \mathbf{k}, \mathbf{l})$ take the values of the cross-terms into account and are also scaled to enhance the contribution of the negative quartets, which would otherwise be swamped by the triplet contribution.

Sheldrick (1990) and Bhat (1990) have introduced the idea of simulated annealing in a phasing procedure. This is a numerical technique, which is used to study systems in thermodynamic equilibrium, and the best-known algorithm is that due to Metropolis, Rosenbluth, Rosenbluth, Teller and Teller (1953). For example, in theoretical studies of the structure of a liquid at some temperature, T, molecules may be placed in random positions in a cubic cell, which is considered to be reproduced in a three-dimensional array to simulate a large (effectively infinite) volume. The potential energy, Φ, of the system is then calculated according to some model of intermolecular potential. A small shift is then made in the position of one of the molecules and the change of potential, $\Delta\Phi$, is calculated. If $\Delta\Phi$ is negative then the new configuration is accepted as one of the ensemble of states being accumulated. If $\Delta\Phi$ is positive then a random number, r, is generated, with uniform probability in the range 0 to 1; if $r < \exp(-\Delta\Phi/kT)$ the new configuration is accepted as one of the ensemble, otherwise the original configuration is restored and added again to the ensemble. By this process an ensemble of states may be assembled where the relative probabilities of different states are those expected from the Boltzmann distribution.

To apply this idea to phase determination we may first consider a centrosymmetric structure. We write (3.71) in the form

$$P_+ = \frac{1}{2} + \frac{1}{2} \tanh[\alpha(\mathbf{h})/2] \qquad (3.119)$$

from which we find

$$\frac{P_+}{P_-} = \exp[-\alpha(\mathbf{h})] \qquad (3.120)$$

where P_+ is the probability that the correct sign is that indicated by $\alpha(\mathbf{h})$. We now introduce a conceptual temperature and express the relative probability of the two states, with positive and negative sign, as $\exp[-\alpha(\mathbf{h})/kT]$. As a modification of the Metropolis algorithm Sheldrick accepted the 'higher energy' state, that with probability P_-, only if the ratio $(P_-/2P_+)$, which must be less than 0.5, was greater than a random number, r, in the range 0 to 1.

The phase-annealing process can also be applied to non-centrosymmetric structures although the algorithm that Sheldrick has found the most efficient departs somewhat from the classical annealing philosophy. In this application, a new estimate, $\phi(\mathbf{h})$ is found for a particular phase from the normal tangent formula, using only three-phase relationships. A magnitude of phase shift is then found from

$$\cos[\varDelta\phi(\mathbf{h})] = \frac{4\alpha(\mathbf{h})/(kT) + \ln(R)}{4\alpha(\mathbf{h})/(kT) - \ln(R)} \tag{3.121}$$

where R is a random number in the range 0 to 1. It will be seen that $\cos[\varDelta\phi(\mathbf{h})]$ is in the range $+1$ to -1 and that if $T=0$ then $\cos[\varDelta\phi(\mathbf{h})]$ will necessarily equal $+1$ so that no change in phase value is made. The phase shift $\varDelta\phi(\mathbf{h})$ is then made to $\phi(\mathbf{h})$ with the sign that best satisfies the available set of negative quartets.

In practical applications of the phase-annealing method Sheldrick started with a large number of random phase sets (1000–10000) and refined for 25 cycles, where a cycle involves considering a sign or phase shift for each structure factor in the system. The initial value of T was estimated by fixing a value of B given by

$$B = \exp\left(-\frac{\alpha_{ran}}{kT}\right), \tag{3.122}$$

where

$$\alpha_{ran} = \langle\alpha(\mathbf{h})_r^2\rangle^{\frac{1}{2}} \tag{3.123}$$

and $\alpha(\mathbf{h})_r^2$ is defined in (3.85). Experimentally it was found that an initial value of $B=3.0$ was satisfactory; this was reduced in each cycle by multiplying T by 0.95. This had the effect of gradually annealing the system into some particular 'low energy', i.e. high probability, state.

The effectiveness of the phase annealing process as found by Sheldrick in some early tests is illustrated in table 3.1. Subsequently it was demonstrated that it could give solutions for structures with up to 180 independent non-hydrogen atoms where more conventional direct methods had failed and it

Table 3.1. *A comparison of the effectiveness of Sheldrick's phase annealing method with the use of his modified tangent formula (3.118). The coded names of the structures are those used in a data bank of test structures. The number of atoms in the unit cell is* N. *Correct solutions were defined as those that could be clearly distinguished by their combined figures of merit and also had all or most atoms as the highest peaks in the corresponding E-map.* M_{MTF} *is the number of solutions for 10 000 trials using the modified tangent formula and* M_A *is the number of solutions for 10 000 trials using the annealing process with an initial value* B = 3.0.

	\multicolumn{7}{c}{Structure code name}						
	LOG	SUOA	PEPI	NEWQB	BHAT	MBH2	HOPS
Space group	$P2_12_12_1$	$P2_12_12_1$	$P2_12_12_1$	P1	Pc	P1	R3
N	108	188	340	124	84	54	243
M_{MTF}	206	6	1	1	26	469	56
M_A	644	50	25	20	97	825	184

has been demonstrated that it could give a solution for a small protein of known structure.

This novel approach is a very useful addition to the armoury of direct methods available to the crystallographer. Direct methods are most useful if all that is available is a structure containing light atoms. However, very often crystal structures contain either heavy atoms, enabling the heavy-atom method to be used, or atoms of moderate atomic number, such as calcium or zinc. The ways in which the presence of moderate-weight atoms may be exploited are considered in chapters 4, 5 and 6.

4

The basics of isomorphous replacement and anomalous scattering

4.1 The isomorphous replacement method

4.1.1 Isomorphous structures

Pairs, or even larger sets, of chemical compounds often occur that are similar and related in a family sense – thus two compounds containing a halogen atom may differ only in having bromine in one structure replaced by iodine in the other. In such a case the crystals produced by these materials may be similar in appearance and when their X-ray diffraction patterns are compared it may be found that the space groups are identical and that both the cell dimensions and patterns of intensity are similar. It may be deduced from this that, to a first approximation, the two structures are identical except that one or more atoms in one structure are replaced, in the same positions, by different kinds of atoms in the other. Such structures are said to be *isomorphous* and a comparison of the intensity data from isomorphous compounds can lead to a determination of their structures. A simple example of a pair of isomorphous compounds is shown in fig. 4.1 where a chlorine atom in one molecule is replaced by a methyl group in the other (Clews and Cochran, 1948). The crystals are needles for both materials, they both have space group $P2_1/a$ with cell dimensions as follows.

	Chlorine compound	Methyl compound
a	16.45 Å	16.43 Å
b	3.85 Å	4.00 Å
c	10.28 Å	10.31 Å
β	108°	109°

A comparison of the diffraction patterns shows that these structures are, indeed, isomorphous.

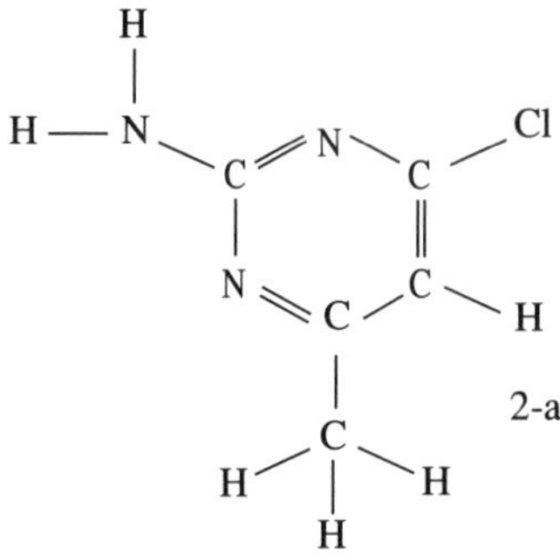

2-amino-4-methyl-6-chloropyrimidine

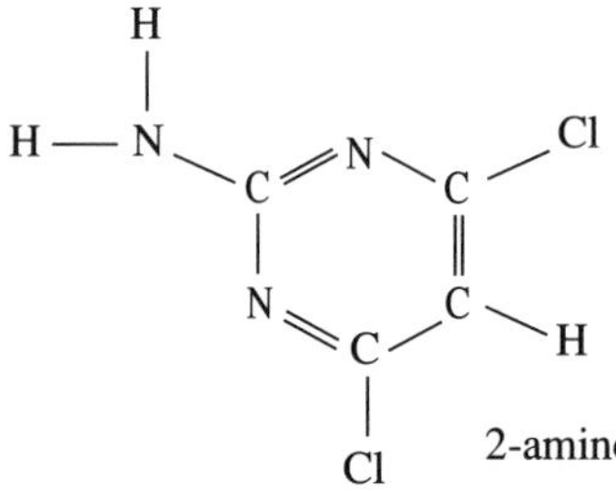

2-amino-4.6-dichloropyrimidine

Fig. 4.1 The isomorphous compounds 2-amino-4-methyl-6-chloropyrimidine and 2-amino-4,6-dichloropyrimidine.

The existence of isomorphous structures has been of the greatest importance for the solution of macromolecular, in particular protein, structures. Protein crystallographers found that, by soaking crystals of a native protein in a solution containing molecules with a heavy-atom component, the crystal structure would sometimes form a heavy-atom derivative with the heavy atom bound, usually in association with a small group of other atoms, at certain sites in the protein. The importance of this procedure is that the derivative is often isomorphous with the native protein and furthermore, in favourable cases, several derivatives with different heavy atoms, perhaps in different locations, can be formed. In fig. 4.2 there are shown precession photographs of the (*hk*0) zones of native lysozyme and of a mercury-containing derivative, lysozyme *p*-chloromercuribenzene sulphonate. Examination of these photographs shows quite clearly that the native protein and the derivative are isomorphous.

The isomorphism of a native protein and a derivative depends on the fact that the protein is a very large structure and is able to accommodate the heavy-atom group by an adjustment that is large in the vicinity of the group but progressively less with greater distance. The general effect of this is that

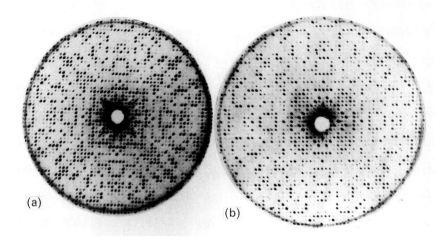

Fig. 4.2 Precession photograph of the (h k 0) zones of (a) native lysozyme, and (b) lysozyme p-chloromercuribenzene sulphonate.

the isomorphism is reasonably good at low resolution but becomes less valid at higher resolution. Typically, phase estimates derived from the isomorphism of a protein and its derivatives may be useful only to, say, 3 Å resolution – but this can often be enough to lead eventually to the complete higher-resolution structure determination via processes of phase extension and refinement (chapter 7).

4.1.2 Isomorphous replacement for centrosymmetric structures

We consider a centrosymmetric structure containing N atoms in the unit cell of which n atoms, of type A, can be replaced by atoms of type B to give an isomorphous structure. For the two structures the structure factors can be written as

$$F_A(\mathbf{h}) = C(\mathbf{h}) + f_A D(\mathbf{h}),$$
$$F_B(\mathbf{h}) = C(\mathbf{h}) + f_B D(\mathbf{h}), \tag{4.1}$$

where $C(\mathbf{h})$ is the contribution to the structure factor of the $N - n$ atoms that are common to both structures and $D(\mathbf{h})$ is given by

$$D(\mathbf{h}) = \sum_{j=1}^{n} \cos(2\pi \mathbf{h} \cdot \mathbf{r}_j). \tag{4.2}$$

If the positions of the isomorphously replaced atoms are known, giving $D(\mathbf{h})$, then it is usually possible to find the signs of $F_A(\mathbf{h})$ and $F_B(\mathbf{h})$. Subtracting the second of the two equations (4.1) from the first to eliminate $C(\mathbf{h})$ gives

$$F_A(\mathbf{h}) - F_B(\mathbf{h}) = (f_A - f_B)D(\mathbf{h}) \qquad (4.3)$$

and if it is known that $|F_A(\mathbf{h})| = 122$, $|F_B(\mathbf{h})| = 107$, $f_A = 16$, $f_B = 22$ and $D(\mathbf{h}) = 2.5$ then the only possible conclusion is that $F_A(\mathbf{h}) = -122$ and $F_B(\mathbf{h}) = -107$. Given that all observations are subject to error, the information may not always be as obvious as in the example just considered but normally the signs of most structure factors, especially the larger ones, will be clearly indicated.

From equation (4.3) it appears that $F_A(\mathbf{h}) - F_B(\mathbf{h})$ is the structure factor for a structure consisting of the n isomorphously replaced atoms with scattering factors $f_A - f_B$. Since $F_A(\mathbf{h})$ and $F_B(\mathbf{h})$ usually have the same sign, especially if the contributions of the isomorphously replaced atoms are not too dominant, a Fourier synthesis computed with coefficients

$$Q(\mathbf{h}) = \left| |F_A(\mathbf{h})| - |F_B(\mathbf{h})| \right|^2 \qquad (4.4)$$

will be a Patterson function corresponding to the n isomorphously replaced atoms. If n is not too large then this is easily interpreted to give the positions of the isomorphously replaced atoms and hence the quantities $D(\mathbf{h})$.

With modern diffraction equipment intensities may be measured with high accuracy and the differences in data between isomorphous structures with isomorphously replaced pairs such as Na–K or O–S may be measured reliably. It should be noted that even the heavier of the pair of isomorphously replaceable atoms may, in itself, be too light to be effective for phasing by the heavy-atom method. For example the addition of a single Hg atom per protein molecule containing 200 amino-acid residues with, say, 40 structured solvent water molecules would give a fractional intensity difference of about 8%. If all that was available was the Hg derivative then, although it might be possible to find the Hg in a Patterson map, phasing from it would be extremely unreliable. However, with both the native protein and derivative available the average 8% difference of intensities can easily be measured, the Hg position can be found more easily from a difference Patterson and phases of many structure factors would be estimated quite reliably. However, it must be remembered that the reliability of the phase estimates would decrease with increasing resolution because of the fall-off in isomorphism.

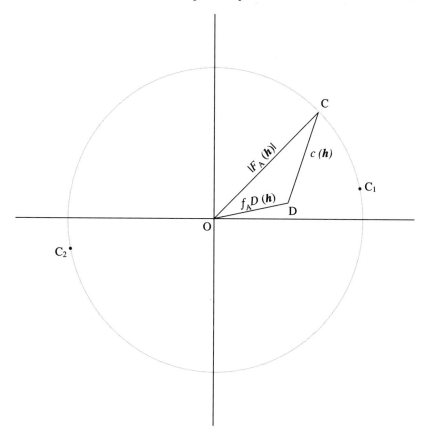

Fig. 4.3 The structure factor for a single isomorphous structure. The phase is undefined.

4.1.3 Single isomorphous replacement (SIR) for non-centrosymmetric structures: the phase ambiguity

Equations (4.1) are equally valid for an isomorphous pair of non-centro-symmetric structures but each term is a complex quantity. The first of the equations (4.1) is shown on an Argand diagram in fig. 4.3. Given that the positions of the isomorphously replaced atoms are known, so that $D(\mathbf{h})$ is known and if $|F_A(\mathbf{h})|$ is also known then we can see from fig. 4.3 that, from the information contained in structure A alone, the phase of $F_A(\mathbf{h})$ is completely unrestricted since the point C may fall anywhere on the circle shown. Of course, Wilson statistics based on the probability distribution of the value of $|C(\mathbf{h})|$ may indicate that C_1 is the most probable position for the

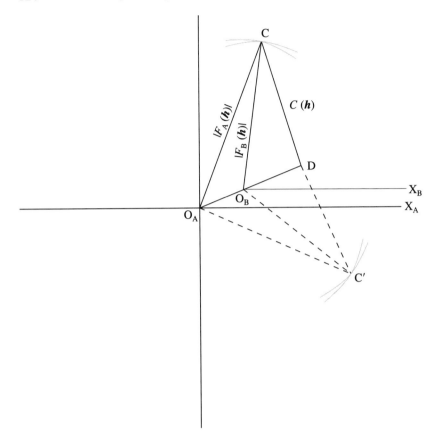

Fig. 4.4 The structure factor for a pair of isomorphous structures. There is a phase ambiguity.

point C and C_2 the least probable. That is, indeed, the basis of the Sim (1960) treatment of the heavy-atom method for the non-centrosymmetric structure where $\alpha(\mathbf{h})$, the phase of $D(\mathbf{h})$, would be assigned to $F(\mathbf{h})$ with some weight. However, the presence of the second structure changes the situation markedly.

Now we refer to the diagram in fig. 4.4; $O_A D$ represents the contribution $f_A D(\mathbf{h})$ and $O_B D$ the contribution $f_B D(\mathbf{h})$. An arc of a circle of radius $|F_A(\mathbf{h})|$ centred on O_A and an arc of radius $|F_B(\mathbf{h})|$ centred on O_B intersect at point C so that DC is a possible contribution $C(\mathbf{h})$. The indicated phase angles for $F_A(\mathbf{h})$ and $F_B(\mathbf{h})$ are $X_A O_A C$ and $X_B O_B C$ respectively. However, it will be noticed that the construction described also gives another intersection of the two arcs at point C′ and this is an equally valid solution, giving possible phase angles $X_A O_A C'$ and $X_B O_B C'$. It can be seen that with a single pair of isomorphous structures, giving the single isomorphous replacement (SIR) method, there is an ambiguity in phase determination for a non-centrosymmetric structure.

4.1.4 SIR for non-centrosymmetric structures: finding the replaced atoms

It is still possible to find vectors between isomorphously replaced atoms from a Fourier synthesis with coefficients $Q(\mathbf{h})$, given by equation (4.4). From fig. 4.4, from geometrical considerations, the magnitude of $(f_A -f_B)D(\mathbf{h})$, $O_A O_B$, must be greater than or equal to the magnitude of $|F_A(\mathbf{h})| - |F_B(\mathbf{h})|$, $O_A C - O_B C$, so that if the latter magnitude is large then so is $D(\mathbf{h})$. The coefficients $Q(\mathbf{h})$ do not accurately represent the coefficients of the Patterson function of the isomorphously replaced structure but there will tend to be some correspondence in that large coefficients, $Q(\mathbf{h})$, used in the map do actually correspond to large Patterson coefficients. Since the number of isomorphously replaced atoms is usually quite small, and the number of reflections available is usually quite large, the problem of finding the vector set of the isomorphously replaced atoms is highly over-determined and they can normally be detected in the map, albeit against a noisy background.

Another direct-methods-based approach is described in §4.2.4 in relation to the use of one-wavelength anomalous scattering data.

4.1.5 SIR for non-centrosymmetric structures: a probabilistic approach to phase estimation

An analytical treatment of the single isomorphous replacement method, taking into account observational errors, was given by Blow and Crick (1959); here we shall look at a simplified treatment of their analysis, which gives the essential result. The model considered by Blow and Crick was of two isomorphous structures that differed in that one contained some extra heavy atoms, which is the situation where there is a native protein and a heavy-atom derivative. We may write

$$F_H(\mathbf{h}) = F_{PH}(\mathbf{h}) - F_P(\mathbf{h}) \qquad (4.5)$$

where $F_P(\mathbf{h})$ and $F_{PH}(\mathbf{h})$ are the structure factors of the native protein and its derivative and $F_H(\mathbf{h})$ is the contribution of the heavy atoms, which is assumed to be known. This equation is the non-centrosymmetric equivalent of (4.3) although, in general, all the terms in it are complex quantities.

There is now required an estimate of the standard error in measuring the difference in magnitudes,

$$\Delta F(\mathbf{h}) = |F_{PH}(\mathbf{h})| - |F_P(\mathbf{h})|.$$

Blow and Crick suggest that for non-centrosymmetric structures where there is a centrosymmetric projection this can be done from the centric reflections alone. If we consider (4.3) in the present notation then, for centric reflections, the ideal result, if there were no errors, is either

$$|F_H(\mathbf{h})| = \left||F_{PH}(\mathbf{h})| - |F_P(\mathbf{h})|\right| \tag{4.6a}$$

or

$$|F_H(\mathbf{h})| = \left||F_{PH}(\mathbf{h})| + |F_P(\mathbf{h})|\right| \tag{4.6b}$$

and when $|F_{PH}(\mathbf{h})|$ and $|F_P(\mathbf{h})|$ are both large result (4.6a) may be either inevitable or very likely. The difference in magnitudes of the two sides of (4.6a) gives a measure of the error in measuring the right-hand-side quantity, assuming that the heavy atoms have been precisely located. The standard error in measuring the difference in the two magnitudes is given by E where

$$E^2 = \overline{\left[|\Delta F(\mathbf{h})| - |F_H(\mathbf{h})|\right]^2}^{\mathbf{h}} \tag{4.7}$$

where the average is taken over all available centrosymmetric reflections.

In fig. 4.5 there is shown the contribution $F_H(\mathbf{h})$, with phase angle $\phi'(\mathbf{h})$, and also $F_P(\mathbf{h})$, which is assumed to make an angle $\Delta\phi(\mathbf{h})$ to $F_H(\mathbf{h})$ to give phase

$$\phi(\mathbf{h}) = \phi'(\mathbf{h}) + \Delta\phi(\mathbf{h}) \tag{4.8}$$

for the native protein. With the measured value of $F_{PH}(\mathbf{h})$ there is a closure error of x. When $F_{PH}(\mathbf{h})$ and $F_P(\mathbf{h})$ are both of large magnitude relative to $F_H(\mathbf{h})$, then $F_{PH}(\mathbf{h})$ and $F_P(\mathbf{h})$ are not far from being parallel so that x can be considered as a consequence of the error in measuring the quantity $\Delta F(\mathbf{h})$, for which we have a known standard error E. If the values of $\Delta F(\mathbf{h})$ have a Gaussian distribution then we may consider the probability that $\phi(\mathbf{h})$ lies between ϕ and $\phi + \mathrm{d}\phi$ to be

$$P(\phi)\,\mathrm{d}\phi = N\exp(-x^2/2E^2)\,\mathrm{d}\phi \tag{4.9}$$

where N is a normalising constant.

However, from fig. 4.4 we see that there are similar probabilities for $\phi'(\mathbf{h}) + \Delta\phi(\mathbf{h})$ and $\phi'(\mathbf{h}) - \Delta\phi(\mathbf{h})$ so that over the range of ϕ from 0 to 2π the total probability will be the sum of two equal Gaussian curves centred on

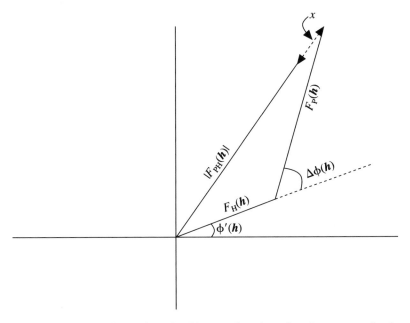

Fig. 4.5 The structure factor for a fixed isomorphously replaced atom contribution and assumed phase for the native protein. The quantity x represents the lack of closure for the measured $|F_P(\mathbf{h})|$ and $|F_{PH}(\mathbf{h})|$.

the two values of ϕ that give zero closure error. This will give a probability curve as shown in fig. 4.6, which must be normalised so that the total area under the curve is unity. If $\Delta\phi$ is small then the two Gaussian curves may merge to give an overall unimodal probability distribution (see fig. 4.7(c)).

4.1.6 SIR for non-centrosymmetric structures: the double-phase synthesis

We have seen in §4.1.3 that for a non-centrosymmetric structure, if there is only one heavy atom derivative, then there is an ambiguity in the phase estimate for the native protein. An obvious way to use the SIR information is to include both possible phases for each reflection in a Fourier calculation. This idea was introduced by Bokhoven, Schoone and Bijvoet (1951) for solving the structure of strychnine sulphate pentahydrate. Approaches along the same lines have been made by Ramachandran and Raman (1959), Raman (1959a), Blow and Rossman (1961) and Kartha (1961). In the basic formulation of this technique a Fourier synthesis is calculated where for the term of index $\mathbf{h}$ the coefficient is

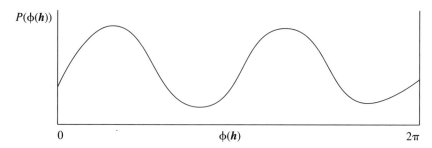

Fig. 4.6 A Blow and Crick probability curve for $\phi(\mathbf{h})$ from a single pair of isomorphous structures.

$$\xi(\mathbf{h}) = \tfrac{1}{2}|F(\mathbf{h})|\{\exp\{i[\phi'(\mathbf{h}) + \Delta\phi(\mathbf{h})]\} + \exp\{i[\phi'(\mathbf{h}) - \Delta\phi(\mathbf{h})]\}\}$$
$$= |F(\mathbf{h})|\cos[\Delta\phi(\mathbf{h})]\exp[i\phi'(\mathbf{h})] \tag{4.10}$$

which is equivalent to using the phase indicated by the isomorphously replaced atoms but with a structure amplitude changed by a factor $\cos[\Delta\phi(\mathbf{h})]$. The rationale of this procedure is that one of the component contributions contained in (4.10) will be correct, within the accuracy limits of the method, and these correct components will systematically build up an image of the structure. On the other hand, the incorrect components will be random in their effects so they will simply contribute a noisy background to the correct image. Blow and Rossman used an improved formula

$$\xi'(\mathbf{h}) = |F(\mathbf{h})|\,\overline{\cos[\Delta\phi(\mathbf{h})]}\exp[i\phi'(\mathbf{h})] \tag{4.11}$$

which took account of the way that experimental errors affected the precision of phase determination by using an average value of $\cos[\Delta\phi(\mathbf{h})]$. For example, if $F_H(\mathbf{h})$ is small then the circles in fig. 4.4 will cut each other at a small angle thus making $\Delta\phi(\mathbf{h})$ poorly defined. This will lead to values of x in equation (4.9) that vary little over wide ranges of $\Delta\phi(\mathbf{h})$, and hence of $\phi(\mathbf{h})$. Regarding the expression (4.9) as a probability associated with $\Delta\phi(\mathbf{h})$ Blow and Rossman used in (4.11)

$$\overline{\cos[\Delta\phi(\mathbf{h})]} = \frac{\displaystyle\int_0^\pi \cos\alpha\,\exp(-x^2/2E^2)\,\mathrm{d}\alpha}{\displaystyle\int_0^\pi \exp(-x^2/2E^2)\,\mathrm{d}\alpha} \tag{4.12}$$

where, for convenience, α has been used on the right-hand side in place of $\Delta\phi(\mathbf{h})$. Blow and Rossman took the expected error in estimating the phase

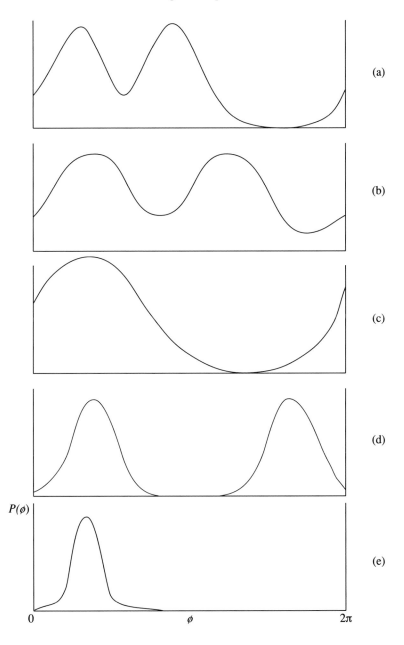

Fig. 4.7(a), (b), (c) and (d) Probability curves for the phase of the native protein from four derivatives. (e) The combined probability curve.

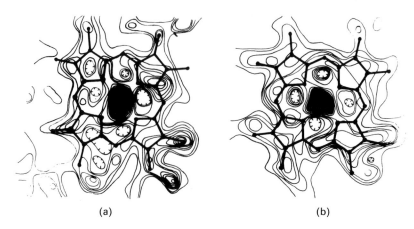

(a)   (b)

Fig. 4.8 The haem group of myoglobin found from (a) single isomorphous replacement by Blow and Rossman, and (b) multiple isomorphous replacement by Kendrew *et al.*

angle as the standard deviation, σ, of the probability distribution (4.9) and they showed that an appropriate weight, or figure of merit, is $\exp(-\sigma^2/2)$, or approximately $\cos \sigma$, for small values of σ.

In fig. 4.8 there is shown a map for the haem section of myoglobin, as given by Blow and Rossman, compared with a map given by Kendrew, Dickerson, Strandberg, Hart, Davies, Phillips and Shore (1960) where the data from five different isomorphous derivatives were employed. While the fit of the haem group to the SIR map is not as good as that obtained by Kendrew and his colleagues it is clear that the SIR method is capable of giving valuable structural information.

The idea that the incorrect phases in (4.10) give rise to a random background breaks down if the heavy atoms have a centrosymmetric distribution. In that case OD in fig. 4.4 will lie along the real axis and the wrong-phase contributions will systematically build up an enantiomorph image of the structure. This makes the task of interpretation much more difficult – if not impossible for a large structure.

4.1.7 *The multiple isomorphous replacement method for non-centrosymmetric structures*

We now consider how to use the information from several isomorphous compounds. The multiple isomorphous replacement (MIR) method is particularly powerful when a number of protein derivatives all have the heavy atoms bound in different sites for then the values of $\phi'(\mathbf{h})$ and $\Delta\phi(\mathbf{h})$ will all be different. The way of combining the information from the various derivatives is to find a probability density function, as in fig. 4.6, for each derivative taken in turn with the native structure. These functions, multiplied together, give an overall probability function. This process is illustrated with a hypothetical example in fig. 4.7 where the sets of information from four isomorphous derivatives are combined. What is seen is that four individual curves, each rather diffuse in nature, can combine to give a well-defined phase estimate.

The MIR method, in terms of what it has contributed to macromolecular structural crystallography, is probably the single most important technique available up to the present time. An early example of its successful use was in the solution of the structure of sperm-whale myoglobin (Bodo, Dintzis, Kendrew and Wyckoff, 1960) with space group $P2_1$. Three isomorphous heavy-atom derivatives were produced: p-chloro-mercuri-benzene sulphonate (PCMBS), mercury diammine ($HgAm_2$) and aurichloride (Au) myoglobin respectively. The x and z heavy-atom coordinates, corresponding to the centrosymmetric projection along y, were found by computing the difference-Patterson map with coefficients $Q(\mathbf{h})$ given by (4.4). There were problems in finding the y coordinates of the heavy atoms, which were eventually found by amalgamating information from four different methods, none of which by itself was very reliable. Two of these methods, one based on an inspection of weighted reciprocal-lattice nets (Bragg, 1958) and the other on a one-dimensional Fourier series using data from the native product and two derivatives (Perutz, 1956), are of limited current interest. A third method, calculating a three-dimensional map with coefficients $Q(\mathbf{h})$, gave a rather confused picture, which did not enable accurate y coordinates to be found. However, the fourth method was rather interesting and utilized the data from some doubly substituted derivatives, namely $PCMBS/HgAm_2$ and PCMBS/Au, prepared by Bodo *et al.* If the structure factors for the native protein, two single derivatives and the double derivative are F, F_{H_1}, F_{H_2} and $F_{H_1H_2}$ respectively then a map with coefficients

$$L = |F_{H_1H_2}|^2 - |F_{H_1}|^2 - |F_{H_2}|^2 + |F|^2$$

will show only the cross vectors between the two kinds of heavy atom. Denoting protein atoms by P the vectors corresponding to the four individual terms in the coefficient are shown below

TERMS	VECTORS					
$\lvert F_{H_1H_2}\rvert^2$	H_1H_1 $\;H_1H_2$	H_2H_2	$P\,H_1$	$P\,H_2$	$P\,P$
$-\lvert F_{H_1}\rvert^2$	$-H_1H_1$		$-P\,H_1$		$-P\,P$
$-\lvert F_{H_2}\rvert^2$		$-H_2H_2$		$-P\,H_2$	$-P\,P$
$+\lvert F\rvert^2$					$P\,P$

RESULTANT	L	H_1H_2

A problem with this process was that of finding a common scale for the different data sets but nevertheless the heavy-atom coordinates were detected with sufficient precision to allow phasing by MIR to 6 Å resolution.

The process of combining information to eliminate SIR ambiguities was done graphically since the Blow and Crick (1959) analytical method had not yet been devised. When the SIR situation involves the native protein and a single derivative the diagram giving the ambiguity appears as in fig. 4.9. The heavy circle, centred on O, has the radius of the native structure amplitude. The heavy-atom contribution to the derivative structure factor is DO (note the direction) and the two possible values for the phase of the native structure factor are XÔC and XÔC'. In fig. 4.10 there is reproduced the figure given by Bodo *et al.* (1960) showing how they combined the information from five derivatives (two doubly substituted) to estimate the phases of the native protein. The direction chosen was that of the point on the heavy circle where all five faint circles approached closely.

From the 6 Å resolution phases Bodo *et al.* were able to define the general form of the molecule and the arrangement of molecules within the crystal; this is shown in fig. 4.11. This great early triumph of macromolecular crystallography is an excellent illustration of the power of the MIR method, achieved at a time when both computers and data-collecting instrumentation were fairly primitive. Further developments in the application of the isomorphous replacement method are given in chapter 5.

4.2 The anomalous scattering method

4.2.1 The phenomenon of anomalous scattering

The coherent scattering of X-rays is by Thomson scattering, the process described in §1.4.1 and §1.4.2. In most situations the frequency of the radiation used is much higher than the natural frequency of any of the atomic electrons and so, by the well-known classical theory of forced

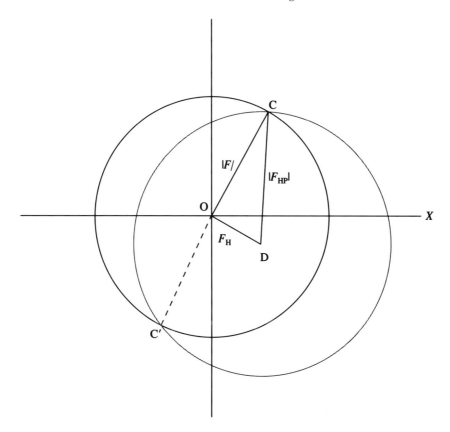

Fig. 4.9 The SIR diagram for a native protein and a heavy-atom derivative.

vibrations, the electrons will oscillate π radius out of phase with the oncoming radiation. This is also the phase shift between the oncoming and scattered radiation but since it is the same for all atoms there is no need to take it into account in the interpretation of diffraction patterns. However, for heavier atoms the inner electrons, particularly those in the K shell, will have very high natural frequencies and the X-radiation used may have similar frequencies. This is the basis of the absorption edges seen in fig. 1.17 where the absorption rises sharply as the frequency of the X-radiation is increased to the value of a natural frequency for one or other of the atomic electrons.

When the oncoming X-radiation has a frequency close to a natural frequency then both the magnitude and the phase of the scattered radiation are different from that expected from the usual Thomson scattering. This is

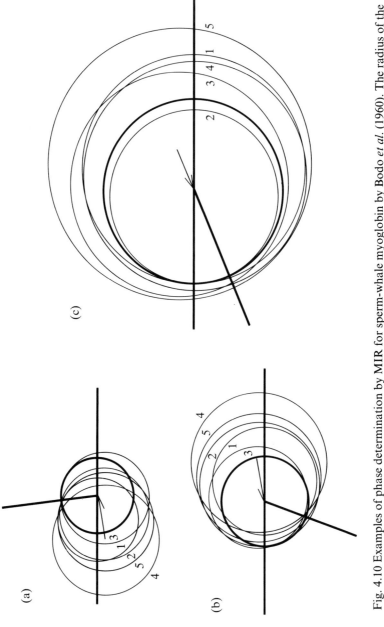

Fig. 4.10 Examples of phase determination by MIR for sperm-whale myoglobin by Bodo *et al.* (1960). The radius of the heavy circle equals the structure amplitude of the native protein. The light circles give the structure amplitudes of 1, PCMBS; 2, HgAm$_2$; 3, Au; 4,, PCMBS/HgAm$_2$; and 5, PCMBS/Au. The heavy line from the centre indicates the final choice of phase angle for the native protein.

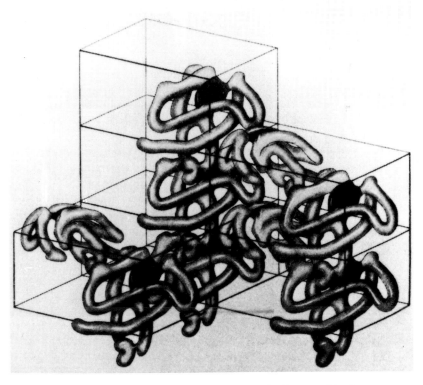

Fig. 4.11 The arrangement of myoglobin molecules in the crystal lattice as seen at 6 Å resolution.

referred to as anomalous scattering and we write the scattering factor in the form

$$f = f^\circ + f' + \mathrm{i}f'' \qquad (4.13)$$

where f° is the normal scattering and the anomalous scattering component is divided into f', which is real and in phase with f° if it is positive, and $\mathrm{i}f''$, an imaginary component with phase $\pi/2$ in advance of f° if f'' is positive. In fig. 4.12 f' and f'' are shown as functions of wavelength near the K absorption edge for iron derived from theory based on quantum mechanics. It will be seen that f' may be positive or negative but is negative in the vicinity of the absorption edge. In addition f'' is always positive but is only non-zero for lower wavelengths (higher frequencies) than the absorption edge. Since the K electrons, which give the anomalous scattering, are so closely bound to the nucleus, scattering from them resembles that from a point atom and it is

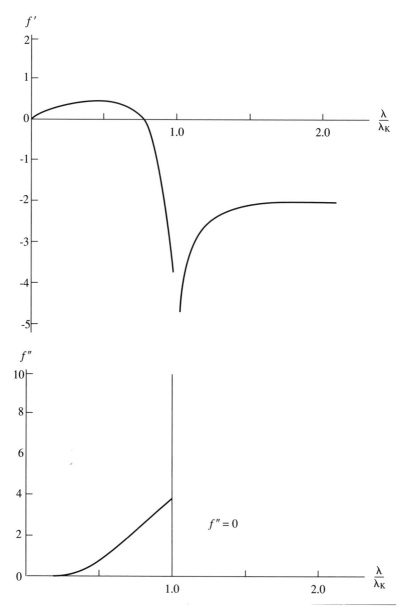

Fig. 4.12 The anomalous scattering components f' and f'' for iron.

usually assumed that values of f' and f'' are constant for all scattering angles.

4.2.2 Anomalous scattering and Friedel's law

When anomalous scattering occurs the structure-factor equation (1.21) may be written as

$$F(\mathbf{h}) = \sum_{j=1}^{N} (f_j^\circ + f_j' + if_j'') \exp(2\pi i \mathbf{h} \cdot \mathbf{r}_j)$$
$$= F^\circ(\mathbf{h}) + F'(\mathbf{h}) + F''(\mathbf{h}), \tag{4.14}$$

where the structure factor components derive from the corresponding scattering factor components and f' and f'' are non-zero for the anomalously scattering atoms. An important consequence of the occurrence of anomalous scattering is that it leads to a breakdown of Friedel's law (1.26). This is illustrated in fig. 4.13, which shows structure factors $F(\mathbf{h})$ and $F(\bar{\mathbf{h}})$ in the complex plane for a structure in which all the atoms but one scatter normally. The contributions of the normally scattering atoms plus the normal component of the anomalously scattering atoms are $F^\circ(\mathbf{h})$ and $F^\circ(\bar{\mathbf{h}})$, which have the same magnitude and opposite phases. The real components of the anomalous scattering are indicated by $F'(\mathbf{h})$ and $F'(\bar{\mathbf{h}})$ and these would preserve Friedel's law since OP and OQ have equal magnitudes and opposite phases. The imaginary components of the anomalous scattering, $F''(\mathbf{h})$ and $F''(\bar{\mathbf{h}})$, each advanced in phase by $\pi/2$ compared with the real part of the anomalous scattering, can be seen to lead to breakdown of Friedel's law.

The departure from Friedel's law has been used to find the absolute configurations of molecules. It is clear from (1.21) that if every atomic position $\mathbf{r}$ were transformed to $-\mathbf{r}$ then the magnitudes of the structure factors would remain unchanged – that is to say that the diffraction pattern from any structure would be the same as that from its enantiomorph. However, this is not true when anomalous scattering occurs and $|F(\mathbf{h})|$ does not equal $|F(\bar{\mathbf{h}})|$. For a solved structure calculated values of $|F(\mathbf{h})|$ and $|F(\bar{\mathbf{h}})|$ can be found and compared with those observed. It is sufficient to measure and observe only a small number of structure factors since what is being looked for is that structure amplitude of the pair of reflections that is the larger. By selecting only those reflections for which the calculated differences of magnitude are large, determinations of absolute configuration can be made with complete certainty.

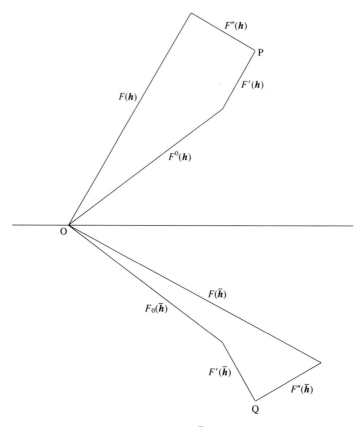

Fig. 4.13 The decomposition of $F(\mathbf{h})$ and $F(\bar{\mathbf{h}})$ into the sum of the normal scattering, the real part of the anomalous scattering and the imaginary part of the anomalous scattering.

4.2.3 *Anomalous scattering and the phase ambiguity*

It is intuitively obvious that there must be important structural information contained in the Friedel differences of structure amplitudes

$$\Delta F(\mathbf{h}) = |F(\mathbf{h})| - |F(\bar{\mathbf{h}})| \qquad (4.15)$$

or of intensities

$$\Delta |F(\mathbf{h})|^2 = |F(\mathbf{h})|^2 - |F(\bar{\mathbf{h}})|^2 \qquad (4.16)$$

in particular concerning the positions of the anomalous scatterers which are responsible for the differences. In principle, if the positions of the anomalous scatterers were known, then it is possible to get phase information, although not unambiguously. This is illustrated in fig. 4.14 which portrays

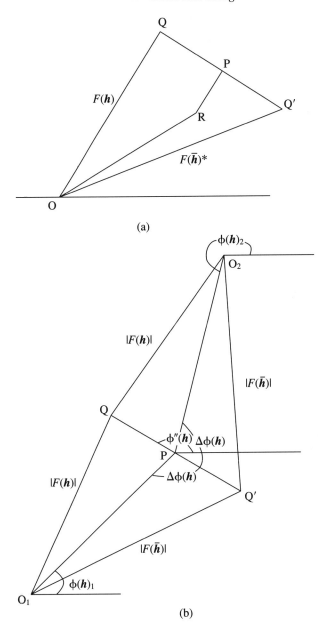

Fig. 4.14(a) The formation of $F(\mathbf{h})$ and $F(\bar{\mathbf{h}})^*$ from their components. (b) Illustrating the phase ambiguity for one-wavelength anomalous scattering. The phase of the non-anomalous component is either $\phi(\mathbf{h})_1$ or $\phi(\mathbf{h})_2$, the values of which are given in (4.17).

the information contained in fig. 4.13 in a more revealing way. As an intermediate representation in fig. 4.14(a) there is shown the result of reflecting the lower part of fig. 4.13 in the real axis. The line OR now represents both $F^{\circ}(\mathbf{h})$ and $F^{\circ}(\bar{\mathbf{h}})^{*}$ and similarly RP represents both $F'(\mathbf{h})$ and $F'(\bar{\mathbf{h}})^{*}$. For the imaginary scattering component, PQ and PQ' represent $F''(\mathbf{h})$ and $F''(\bar{\mathbf{h}})$ respectively. The lines OQ and OQ' correspond to $F(\mathbf{h})$ and $F(\bar{\mathbf{h}})^{*}$, the magnitudes of which constitute the observed data.

We turn to fig. 4.14(b), which is best interpreted in relation to fig. 4.14(a). If the positions of the anomalous scatterers are known then the imaginary parts of their anomalous contributions are known both in magnitude and phase so that PQ and PQ' are known. To find the point O we construct on the base QQ' a triangle for which the other two sides have lengths $|F(\mathbf{h})|$ and $|F(\bar{\mathbf{h}})|$ but this can be done in two ways leading to the points O_1 and O_2. The contribution of the total real part of the scattering, including that from the anomalous scatterers, is either O_1P or O_2P and the phase associated with this contribution, knowledge of which for each reflection would enable us to solve the structure, is

$$\phi(\mathbf{h}) = \phi''(\mathbf{h}) \pm \Delta\phi(\mathbf{h}). \tag{4.17}$$

The resolution of this phase ambiguity leads to a solution of the structure and the various ways in which this can be done are dealt with in chapter 6.

4.2.4 Determination of positions of the anomalous scatterers

Multiplying (4.14) by its complex conjugate, taking account of the fact that for each term involving (i, j) there is another involving (j, i), gives

$$|F(\mathbf{h})|^2 = \sum_{i=1}^{N} \sum_{j=1}^{N} \left[(f_i + f_i')(f_j + f_j') + f_i'' f_j'' \right] \cos\left[2\pi i \mathbf{h} \cdot (\mathbf{r}_i - \mathbf{r}_j) \right]$$
$$- \sum_{i=1}^{N} \sum_{j=1}^{N} \left[(f_j + f_j')f_i'' - (f_i + f_i')f_j'' \right] \sin\left[2\pi i \mathbf{h} \cdot (\mathbf{r}_i - \mathbf{r}_j) \right]. \tag{4.18}$$

From this we find

$$\tfrac{1}{2}\left(|F(\mathbf{h})|^2 + |F(\bar{\mathbf{h}})|^2\right) = \sum_{i=1}^{N} \sum_{j=1}^{N} \left[(f_i + f_i')(f_j + f_j') + f_i'' f_j'' \right] \cos\left[2\pi i \mathbf{h} \cdot (\mathbf{r}_i - \mathbf{r}_j) \right].$$
$$\tag{4.19}$$

The left-hand side of (4.19) can be used as the Fourier coefficient for a Patterson function, which will reveal the vectors between the anomalous scatterers if they are heavy atoms in the normal sense.

If the anomalously scattering atoms are not very heavy, for example iron, then it is better to use the values of the anomalous differences, $\Delta F(\mathbf{h})$, as given by (4.15). From fig. 4.14(b), where there is a triangle with sides equal to $2|F''(\mathbf{h})|$, $|F(\mathbf{h})|$ and $|F(\bar{\mathbf{h}})|$ it is evident that

$$2|F''(\mathbf{h})| \geq |\Delta F(\mathbf{h})| = \left||F(\mathbf{h})| - |F(\bar{\mathbf{h}})|\right|. \qquad (4.20)$$

From (4.14)

$$|F''(\mathbf{h})| = \left|\sum_{j=1}^{M} f_j'' \exp(2\pi i \mathbf{h} \cdot \mathbf{r}_j)\right| \qquad (4.21)$$

where there are M anomalous scatterers in the unit cell. From (4.20) and (4.21) it is clear that if $|\Delta F(\mathbf{h})|$ is large then so is the structure factor of the sub-structure consisting just of the anomalously scattering atoms. In this way there can be identified some of the large structure factors of the sub-structure but not all of them. It is possible to have a large value of $|F''(\mathbf{h})|$ but a small or even zero value of $|\Delta F(\mathbf{h})|$; this corresponds to having an isosceles triangle on a long base. However, the presence of a substantial subset of large anomalous-difference magnitudes will enable the positions of the anomalous scatterers, which are usually few in number, to be found.

If the number of anomalous scatterers is ten or less then a Patterson function with $|\Delta F(\mathbf{h})|$ as coefficients can usually be interpreted to give their positions. For a larger number of anomalous scatterers Mukherjee, Helliwell and Main (1989) have suggested use of the $|\Delta F(\mathbf{h})|$ values as input in MULTAN to determine directly their positions. In addition they took a number of the smallest values of $|\Delta F(\mathbf{h})|$ to use in a PSIZERO figure-of-merit test although these could not reliably indicate small values of $|F''(\mathbf{h})|$. Mukherjee *et al.* showed that, for three fairly large proteins for which anomalous-scattering data were available and also for one small structure, it was possible to determine the positions of the anomalous scatterers. Details of the structures to which they applied MULTAN are shown in table 4.1 with the results that they obtained in table 4.2. In each case the E-map corresponding to the largest value of CFOM revealed the anomalous scatterers although sometimes with large positional errors – up to 5 Å in the case of pea lectin. All these structures had a small number of anomalous scattering sites but it seems likely that it would also be successful with a larger number of sites as might occur with genetically engineered proteins.

With the positions of the anomalous scatterers known then it seems that phasing is possible but only if the phase ambiguity evident in fig. 4.14(b) can be resolved. How this may be achieved is dealt with in chapter 6.

Table 4.1. *Data for the structures used in the MULTAN determination of the positions of anomalous scatterers*

	Cytochrome c_4	Pea lectin	Hg derivative of α-amylase	Small molecule $C_{10}H_{11}ClFNO$	
Space group	$P6_122$	$P2_12_12_1$	$C222_1$	$P2_12_12_1$	
Unit cell a (Å)	62.4	50.8	81.1	5.42	
b (Å)	62.4	61.6	98.3	12.03	
c (Å)	174.2	137.4	138.0	15.84	
Molecular weight	19 000	49 000	45 000	216	
Wavelength (Å)	1.74	1.86	1.542	1.542	0.7107
Resolution (Å)	3.0	3.0	5.5	1.0	1.0
Number of reflections in map	400	400	300	414	338
Number of phase sets generated	28	200	100	50	50

4.3 Assessing the quality of solutions for proteins

The techniques of isomorphous replacement and anomalous scattering have their major applications in the solution of protein structures and in chapters 5 and 6 the considerable successes of these methods will become apparent. In the solution of an unknown structure the only measure of the quality of a set of phases is the interpretability of the resultant electron-density map in terms of a structural model and the success of subsequent phase extension and refinement processes (chapter 7). However, in the development of methods, which is usually done with known trial structures, it is useful to have some measure of attainment so that the effectiveness of methods can be judged both in an absolute way and relative to each other. Since the critical stage in protein crystallography is the interpretability of an electron-density map, a common measure of quality is the map correlation coefficient defined as

$$r = \frac{\overline{\rho_e \rho_t} - \overline{\rho_e}\,\overline{\rho_t}}{\sigma_{\rho_e} \sigma_{\rho_t}} \qquad (4.22)$$

where ρ_e and ρ_t are the electron densities derived from estimated phases and true phases respectively and σ_{ρ_e} and σ_{ρ_t} are their standard deviations. These quantities are usually found from the values at the grid points at which the densities are calculated. The quantity r is a standard linear correlation coefficient which must satisfy the relationship $-1 \leq r \leq 1$. In practice a

Table 4.2. *Results from the* E-*map with the highest CFOM. The next highest peak, other than those showing the positions of the anomalous scatterers, are also given.*

Structure	Peak height	Calculated positions			Atom	True positions		
		x	y	z		x	y	z
Cytochrome	1318	0.454	−0.003	0.316	Fe_1	0.457	0.003	0.318
c_4	1141	0.653	0.102	0.082	Fe_2	0.656	0.102	0.082
	307	0.826	0.096	0.851				
Pea lectin	868	0.899	0.102	0.101	Mn_1	0.89	0.14	0.13
	814	0.600	0.147	0.259	Mn_2	0.66	0.10	0.29
	694	0.865	0.066	0.806				
α-amylase	3090	0.109	0.230	0.529	Hg	0.140	0.229	0.528
Hg derivative	741	0.282	0.172	0.619				
$C_{10}H_{11}ClFNO$	2828	0.807	0.288	0.770	Cl	0.798	0.287	0.773
(Cu Kα data)	837	0.191	0.137	0.543				
$C_{10}H_{11}ClFNO$	2855	0.857	0.299	0.808	Cl	0.798	0.287	0.773
(Mo Kα data)	788	0.799	0.166	0.754				

value of r greater than about 0.5 usually means that the map can readily be interpreted. However, in favourable cases values of 0.4, or even less, can be informative.

We shall be using r as a measure of quality when methods of using anomalous scattering data are described in chapter 6.

5

Further aspects of the isomorphous replacement method

5.1 Introduction

The crystal structure analysis of a protein would be a routine procedure provided that two or more heavy-atom derivatives with good isomorphism to the native protein were available (chapter 4). Unfortunately, in practice this is often not the case. Usually there will be little difficulty in preparing one heavy-atom derivative that is isomorphous with the native protein. However, finding a second isomorphous derivative may not be straightforward so that the use of single isomorphous replacement (SIR) data is preferable if there is some way to resolve the intrinsic ambiguity of the method in the non-centrosymmetric case (§4.1.3). The double-phase method (§4.1.6) is one way of doing this but it usually gives a rather noisy map. There is now described a noise-filtering technique which can be used to develop better information in such a situation.

5.2 Resolving the SIR phase ambiguity in real space: Wang's solvent-flattening method

Protein structures are characterized by having large contiguous solvent regions surrounding other regions of somewhat higher average density within which the protein exists. The contrast between the ordered structure (protein, sometimes with some solvent molecules) and background (disordered solvent) is much less than for small-molecule structures and this is one of the reasons, additional to other factors including their size and complexity, which make protein structures difficult to solve.

The first critical step in Wang's method (Wang, 1981, 1985) is to define the molecular boundary from a noisy electron density map. Following that, the densities inside the protein envelope are raised by a constant value and then densities lower than a certain value are removed. Outside the protein region, the density is smoothed to a constant level. By this process some of the major noise in the Fourier map is filtered and a rough structural model is constructed, which consists of the ordered (protein) and disordered (solvent) regions. This can then be used for resolving the phase ambiguity. In practice, Wang's procedure is divided into seven steps.

1. Calculate the phase and figure of merit for each reflection from the SIR data (see Blow and Crick, 1959 (§4.1.5) and Hendrickson and Lattman, 1970).
2. Calculate a Fourier map.
3. Locate the molecular boundary from the Fourier map. An automatic process was devised for this purpose, which is based on the assumption that average density within the protein region should be higher than that of the solvent region. A new map $\rho'(\mathbf{r})$ is constructed such that the density ρ'_j at the grid point j is proportional to the weighted sum of the densities ρ_i (above a certain background value) within a sphere of radius R centred on that grid point in the original map $\rho(\mathbf{r})$, i.e.

$$\rho'_j = k \sum_i w_i \rho_i, \, w_i = \begin{cases} 1 - r_{ij}/R & \text{if } r_{ij} \le R \text{ and } \rho_i > 0 \\ 0 & \text{if } r_{ij} > R \text{ or } \rho_i \le 0, \end{cases} \quad (5.1)$$

where k is a constant and r_{ij} is the distance between grid points i and j. Note that ρ does not include the contribution from $F(000)$. The best radius of summation, R has been found to vary with the resolution of the data, being 9 Å and 12 Å for 3 Å and 6 Å data respectively. The resulting new map has large fairly homogeneous connected regions of relatively high and low density. The molecular boundary is then determined by setting a threshold chosen so that the total volume (number of grid points) with density lower than the threshold corresponds to the volume known to contain solvent.

4. Filter errors in real space. Before filtering, a constant density, ρ^C, is added to every grid point of the map such that the average solvent density after the addition will satisfy the following constraint:

$$\frac{\rho^C + \langle \rho_{sol} \rangle}{\rho^C + \rho_{max}} = S, \quad (5.2)$$

where $\langle \rho_{sol} \rangle$ and ρ_{max} are respectively the average solvent density and the maximum protein density in the Fourier map, and S is a constant to be determined empirically. For Bence Jones protein Rhe grown from a solution containing 2M $(NH_4)_2SO_4$ (Wang, Yoo and Sax, 1979), the value of S was found to be 0.06 and 0.25 respectively for 3 Å and 6 Å data. The noise filtering on the map after adding ρ^C is done as follows. Inside the protein envelope all the negative densities are replaced with zero while within the solvent region the density is replaced with $\rho^C + \langle \rho_{sol} \rangle$.

5. Calculate structure factors by Fourier transforming the map resulting from step 4.

6. Filter errors in reciprocal space. This is done by multiplying the original bimodal SIR phase probability distribution with a Sim distribution (2.44) centred on the phases found in step 5.

7. Calculate new phases and figure of merit and then go to step 2 and repeat the process. Note that step 3 does not have to be repeated in each cycle.

Wang's method has been applied to solve a number of originally unknown protein structures and has been proved to be very efficient. The success, however, relies on two conditions: (i) there must be a sufficiently large amount of solvent in the crystal; (ii) the arrangement of the additional or isomorphously replaced atoms should not possess symmetry higher than that of the structure as a whole. Leslie (1987) proposed a reciprocal-space algorithm for locating the molecular envelope, which speeds up the most time-consuming step in Wang's process. Zhang and Main (1990a, b) incorporated Wang's method into a more general approach, which also includes histogram matching, which involves modifying the density in the protein region so as to create a distribution of electron density in agreement with some expected distribution. Further details of this improved procedure are given in chapter 7.

5.3 Resolving the ambiguity in reciprocal space: a general review

The phase ambiguity in reciprocal space arising from the SIR technique, as seen in fig. 5.1, may be written as

$$\phi(\mathbf{h}) = \phi_R(\mathbf{h}) \pm |\varDelta\phi(\mathbf{h})|. \tag{5.3}$$

The problem is to determine the sign with which to apply $|\varDelta\phi|$ for each reflection. Early attempts were made to use direct methods for breaking the phase ambiguity with Coulter (1965) suggesting the use of the tangent formula (3.60). The starting reflections for the process were those for which the two possible SIR phases were fairly close to each other so that the average would have only a small error. Fan (1965b) and similarly Karle (1966) suggested the use of a 'component relation', which relates the real and imaginary parts of structure factors. If the isomorphously replaced atoms have a centrosymmetric arrangement and the origin of the unit cell is

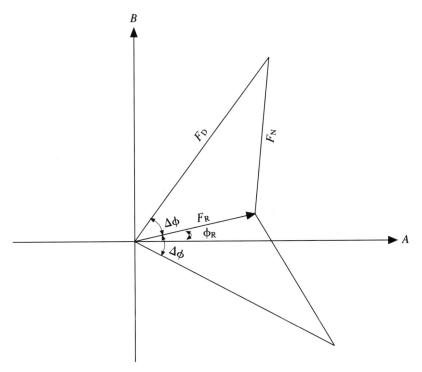

Fig. 5.1 Phase ambiguity arising from the single isomorphous replacement method.

chosen at the inverse centre, then for each reflection we know the real part and the magnitude of the imaginary part of the structure factor (see fig. 5.1). The component relation can then be used to derive the signs of the imaginary parts and hence to resolve the phase ambiguity. Hendrickson (1971) tried to break the ambiguity by multiplying the bimodal SIR distribution with the distribution of three-phase structure invariants given by Cochran (1955).

During the 1980s there was considerable progress in developing new approaches. Probabilistic formulae for integrating direct methods with SIR data were derived by Hauptman (1982a) and further improvements were made by Fortier, Moore and Fraser (1985), Klop, Krabbendam and Kroon (1987) and Hao and Fan (1988). Algebraic equations were given by Karle (1983) and also by Langs (1986) for the estimation of three-phase structure invariants and individual phases using SIR data. A procedure based on the work of Fan (1965b) was proposed, which combined direct methods with macromolecular methods to break the SIR phase ambiguity (Fan, Han, Qian and Yao, 1984; Fan, Han and Qian, 1984; Fan and Gu, 1985; Yao and

Fan, 1985). Actually this is the only direct-method procedure, up to the present, which has been successful in resolving the SIR phase ambiguity in a test using experimental protein diffraction data (Fan and Wang, 1988, unpublished work). Although none of the direct methods have been as successful as Wang's method described in §5.2, the combination of the two types of method may provide better results than either of them alone.

Some of the more recent developments are now examined in greater detail.

5.3.1 Algebraic methods

Based on the mathematical and physical characteristics of SIR, Karle (1983) proposed a simple rule, which enables the selection of three-phase structure invariants having values close to zero or π. He also derived a simple formula, which permits the estimation of the cosine of three-phase invariants from the diffraction data. This provides a starting point for the solution of the SIR phase-ambiguity problem.

For SIR data from a protein and a derivative we have the equation

$$F_N(\mathbf{h}) = F_D(\mathbf{h}) - F_R(\mathbf{h}), \tag{5.4}$$

where $F_N(\mathbf{h})$ is the structure factor of the native protein, $F_D(\mathbf{h})$ is that of the derivative and $F_R(\mathbf{h})$ is the contribution of the isomorphously replaced atoms. By rearranging (5.4)

$$F_R(\mathbf{h}) = F_D(\mathbf{h}) - F_N(\mathbf{h}). \tag{5.5}$$

From (5.5), and by considering the nature of isomorphous replacement, Karle made the following observations.

1. The largest differences $||F_D(\mathbf{h})| - |F_N(\mathbf{h})||$ are associated with the largest $|F_R(\mathbf{h})|$.
2. Three-phase structure invariants associated with the largest $|F_R(-\mathbf{h})F_R(\mathbf{h}')F_R(\mathbf{h}-\mathbf{h}')|$ are expected to be approximately zero since the isomorphously replaced structure is usually a simple one for which three-phase relationships should hold well.
3. For larger values of $||F_D(\mathbf{h})| - |F_N(\mathbf{h})||$, the phase $\phi_N(\mathbf{h})$ will differ little from the phase $\phi_D(\mathbf{h})$.

From (5.5) it follows that

$$F_R(-\mathbf{h})F_R(\mathbf{h}')F_R(\mathbf{h}-\mathbf{h}') = \left[F_D(-\mathbf{h}) - F_N(-\mathbf{h})\right] \cdot \left[F_D(\mathbf{h}') - F_N(\mathbf{h}')\right] \cdot \left[F_D(\mathbf{h}-\mathbf{h}') - F_N(\mathbf{h}-\mathbf{h}')\right] \tag{5.6}$$

which can be expanded to give

$$|F_R(-\mathbf{h})F_R(\mathbf{h}')F_R(\mathbf{h}-\mathbf{h}')| \exp\{i[\phi_R(-\mathbf{h})+\phi_R(\mathbf{h}')+\phi_R(\mathbf{h}-\mathbf{h}')]\} =$$
$$|F_D(-\mathbf{h})F_D(\mathbf{h}')F_D(\mathbf{h}-\mathbf{h}')| \exp\{i[\phi_D(-\mathbf{h})+\phi_D(\mathbf{h}')+\phi_D(\mathbf{h}-\mathbf{h}')]\}$$
$$-|F_D(-\mathbf{h})F_N(\mathbf{h}')F_D(\mathbf{h}-\mathbf{h}')| \exp\{i[\phi_D(-\mathbf{h})+\phi_N(\mathbf{h}')+\phi_D(\mathbf{h}-\mathbf{h}')]\}$$
$$-|F_D(-\mathbf{h})F_D(\mathbf{h}')F_N(\mathbf{h}-\mathbf{h}')| \exp\{i[\phi_D(-\mathbf{h})+\phi_D(\mathbf{h}')+\phi_N(\mathbf{h}-\mathbf{h}')]\}$$
$$+|F_D(-\mathbf{h})F_N(\mathbf{h}')F_N(\mathbf{h}-\mathbf{h}')| \exp\{i[\phi_D(-\mathbf{h})+\phi_N(\mathbf{h}')+\phi_N(\mathbf{h}-\mathbf{h}')]\}$$
$$-|F_N(-\mathbf{h})F_D(\mathbf{h}')F_D(\mathbf{h}-\mathbf{h}')| \exp\{i[\phi_N(-\mathbf{h})+\phi_D(\mathbf{h}')+\phi_D(\mathbf{h}-\mathbf{h}')]\}$$
$$+|F_N(-\mathbf{h})F_N(\mathbf{h}')F_D(\mathbf{h}-\mathbf{h}')| \exp\{i[\phi_N(-\mathbf{h})+\phi_N(\mathbf{h}')+\phi_D(\mathbf{h}-\mathbf{h}')]\}$$
$$+|F_N(-\mathbf{h})F_D(\mathbf{h}')F_N(\mathbf{h}-\mathbf{h}')| \exp\{i[\phi_N(-\mathbf{h})+\phi_D(\mathbf{h}')+\phi_N(\mathbf{h}-\mathbf{h}')]\}$$
$$-|F_N(-\mathbf{h})F_N(\mathbf{h}')F_N(\mathbf{h}-\mathbf{h}')| \exp\{i[\phi_N(-\mathbf{h})+\phi_N(\mathbf{h}')+\phi_N(\mathbf{h}-\mathbf{h}')]\}.$$

(5.7)

On the basis of Karle's observation 3, we have to good approximation

$$|F_R(-\mathbf{h})F_R(\mathbf{h}')F_R(\mathbf{h}-\mathbf{h}')| \exp\{i[\phi_R(-\mathbf{h})+\phi_R(\mathbf{h}')+\phi_R(\mathbf{h}-\mathbf{h}')]\}$$
$$\simeq [|F_D(-\mathbf{h})|-|F_N(-\mathbf{h})|] \cdot [|F_D(\mathbf{h}')|-|F_N(\mathbf{h}')|] \cdot [|F_D(\mathbf{h}-\mathbf{h}')|$$
$$-|F_N(\mathbf{h}-\mathbf{h}')|] \exp[i\langle\phi(-\mathbf{h})+\phi(\mathbf{h}')+\phi(\mathbf{h}-\mathbf{h}')\rangle]$$

(5.8)

where $\langle\phi(-\mathbf{h})+\phi(\mathbf{h}')+\phi(\mathbf{h}-\mathbf{h}')\rangle$ represents the average of the eight three-phase invariants on the right-hand side of (5.7). According to Karle's observations 1 and 2, the left-hand side of (5.8) approximates to a real positive number. Therefore, if the triple product on the right-hand side of (5.8) is composed of large magnitude differences, an examination of this equation leads to the following rule:

$$\langle\phi(-\mathbf{h})+\phi(\mathbf{h}')+\phi(\mathbf{h}-\mathbf{h}')\rangle \simeq 0$$

(5.9a)

if

$$[|F_D(-\mathbf{h})|-|F_N(-\mathbf{h})|][|F_D(\mathbf{h}')|-|F_N(\mathbf{h}')|][|F_D(\mathbf{h}-\mathbf{h}')|-|F_N(\mathbf{h}-\mathbf{h}')|]$$

is large in magnitude and positive or

$$\langle[\phi(-\mathbf{h})+\phi(\mathbf{h}')+\phi(\mathbf{h}-\mathbf{h}')]\rangle \simeq \pi$$

(5.9b)

if

$$[|F_D(-\mathbf{h})|-|F_N(-\mathbf{h})|][|F_D(\mathbf{h}')|-|F_N(\mathbf{h}')|][|F_D(\mathbf{h}-\mathbf{h}')|-|F_N(\mathbf{h}-\mathbf{h}')|]$$

is large in magnitude and negative.

Despite its simplicity and its ability to give reliable estimates (0 or π) for a large number of three-phase structure invariants, there is a pitfall in Karle's procedure. Inherent in the idea is that $\phi_D(\mathbf{h})$ is assumed to be approximately equal to $\phi_N(\mathbf{h})$, the implication of which is that $\Delta\phi(\mathbf{h})$ of (5.3) equals zero or π. Consequently the deduced phase for most reflections turns out to be close to the averaged SIR phase, i.e. either $\phi_R(\mathbf{h})$ or $\phi_R(\mathbf{h}) + \pi$. This leaves the ambiguity essentially unresolved. The problem was detected by Xu, Yang, Furey, Sax, Rose and Wang (1984) when they were testing Karle's method with the experimental Au SIR data of Bence Jones protein Rhe (Wang, Yoo and Sax, 1979). They found that the resulting Karle phases were much closer to the averaged SIR phases than to the protein phases.

Algebraic equations have also been derived making use of the known positions of the isomorphously replaced atoms (Langs, 1986). The phase of reflections from the derivative is expressed as

$$\phi_D(\mathbf{h}) = \phi_R(\mathbf{h}) + \tan^{-1}(S_h/G_h) \qquad (5.10)$$

where

$$S_h = 2|F_D(\mathbf{h})F_R(\mathbf{h})|\sin[\phi_D(\mathbf{h}) - \phi_R(\mathbf{h})] \equiv A_h\sin[\Delta\phi(\mathbf{h})]$$
$$G_h = 2|F_D(\mathbf{h})F_R(\mathbf{h})|\cos[\phi_D(\mathbf{h}) - \phi_R(\mathbf{h})] \equiv A_h\cos[\Delta\phi(\mathbf{h})].$$

Since the sign of $\Delta\phi(\mathbf{h})$ could not be found directly from SIR data, in practice approximate values of S_h and G_h are derived as follows. We write the phase identity

$$[\phi_D(\mathbf{h}) - \phi_R(\mathbf{h})] + \Phi_{h,h'} = [\phi_D(\mathbf{h}') - \phi_R(\mathbf{h}')] + [\phi_D(\mathbf{h} - \mathbf{h}') - \phi_R(\mathbf{h} - \mathbf{h}')] + \Psi_{h,h'}$$
$$(5.11)$$

or

$$\Delta\phi(\mathbf{h}) + \Phi_{h,h'} = \Delta\phi(\mathbf{h}') + \Delta\phi(\mathbf{h} - \mathbf{h}') + \Psi_{h,h'} \qquad (5.12)$$

where

$$\Phi_{h,h'} = \phi_D(-\mathbf{h}) + \phi_D(\mathbf{h}') + \phi_D(\mathbf{h} - \mathbf{h}'),$$
$$\Psi_{h,h'} = \phi_R(-\mathbf{h}) + \phi_R(\mathbf{h}') + \phi_R(\mathbf{h} - \mathbf{h}').$$

By taking the sine of both sides of (5.12) and averaging over all $\mathbf{h}'$, we have

$$S_h\langle A_{h'}A_{h-h'}\cos(\Phi_{h,h'})\rangle_{h'} + G_h\langle A_{h'}A_{h-h'}\sin(\Phi_{h,h'})\rangle_{h'} =$$
$$A_h\langle(G_{h'}S_{h-h'} + S_{h'}G_{h-h'})\cos(\Psi_{h,h'}) + (G_{h'}G_{h-h'} - S_{h'}S_{h-h'})\sin(\Psi_{h,h'})\rangle_{h'}.$$
$$(5.13)$$

Similarly we have from the average of the cosine expansion

$$G_h\langle A_{h'}A_{h-h'}\cos(\Phi_{h,h'})\rangle_{h'} - S_h\langle A_{h'}A_{h-h'}\sin(\Phi_{h,h'})\rangle_{h'} =$$
$$A_h\langle(G_{h'}G_{h-h'} - S_{h'}S_{h-h'})\cos(\Psi_{h,h'}) - (G_{h'}S_{h-h'} + S_{h'}G_{h-h'})\sin(\Psi_{h,h'})\rangle_{h'}.$$
$$(5.14)$$

With the approximation $\langle A_{h'}A_{h-h'}\sin(\Phi_{h,h'})\rangle_{h'} \simeq 0$, (5.13) and (5.14) become respectively

$$S_h \simeq A_h\langle (G_{h'}S_{h-h'} + S_{h'}G_{h-h'})\cos(\Psi_{h,h'}) + (G_{h'}G_{h-h'}$$
$$- S_{h'}S_{h-h'})\sin(\Psi_{h,h'})\rangle_{h'}/\langle A_{h'}A_{h-h'}\cos(\Phi_{h,h'})\rangle_{h'}. \quad (5.15)$$

$$G_h \simeq A_h\langle (G_{h'}G_{h-h'} - S_{h'}S_{h-h'})\cos(\Psi_{h,h'}) - (G_{h'}S_{h-h'}$$
$$+ S_{h'}G_{h-h'})\sin(\Psi_{h,h'})\rangle_{h'}/\langle A_{h'}A_{h-h'}\cos(\Phi_{h,h'})\rangle_{h'}. \quad (5.16)$$

Equations (5.15), (5.16) and (5.10) are then used to calculate values of $\phi_D(\mathbf{h})$ by an iterative procedure. Initial estimates of various $S_{h'}$ and $S_{h-h'}$ to insert into the right-hand sides of (5.15) and (5.16) may be obtained by taking just those parts of (5.15) and (5.16) that are of known value, giving

$$S'_h = A_h\langle G_{h'}G_{h-h'}\sin(\Psi_{h,h'})\rangle_{h'}, \quad (5.17)$$
$$G'_h = A_h\langle G_{h'}G_{h-h'}\cos(\Psi_{h,h'})\rangle_{h'}, \quad (5.18)$$
$$S_h \simeq A_h\sin[\tan^{-1}(S'_h/G'_h)]. \quad (5.19)$$

Langs' equations are better than Karle's in the sense that

(i) estimation is made directly of individual phases rather than of three-phase invariants; and
(ii) there is no need to assume that $\phi_D(\mathbf{h}) \simeq \phi_N(\mathbf{h})$.

Test calculations (Langs, 1986) showed that the procedure works reasonably well for a small molecular structure but not as well for a protein such as insulin even with error-free calculated SIR data.

5.3.2 Integration of direct methods with SIR data

Probability formulae integrating direct methods with SIR data were derived by Hauptman (1982a). Given a pair of isomorphous structures, we denote the two sets of structure factors by

$$E_h = |E_h|\exp(i\phi_h) = \frac{1}{\alpha_{20}^{1/2}}\sum_{j=1}^{N} f_j\exp(2\pi i\mathbf{h}\cdot\mathbf{r}_j), \quad (5.20)$$

$$G_h = |G_h|\exp(i\psi_h) = \frac{1}{\alpha_{02}^{1/2}}\sum_{j=1}^{N} g_j\exp(2\pi i\mathbf{h}\cdot\mathbf{r}_j), \quad (5.21)$$

where

$$\alpha_{mn} = \sum_{j=1}^{N} f_j^m g_j^n.$$

The atomic scattering factors are f_j for one of the isomorphous structures and g_j for the other. Accordingly there will be four kinds of three-phase structure invariants:

$$
\begin{aligned}
\omega_0 &= \phi_{-h} + \phi_{h'} + \phi_{h-h'}, \\
\omega_1 &= \phi_{-h} + \phi_{h'} + \psi_{h-h'}, \\
\omega_2 &= \phi_{-h} + \psi_{h'} + \psi_{h-h'}, \\
\omega_3 &= \psi_{-h} + \psi_{h'} + \psi_{h-h'}.
\end{aligned}
\tag{5.22}
$$

The first neighbourhood of each kind of three-phase structure invariant is defined to consist of the six magnitudes $|E_h|$, $|E_{h'}|$, $|E_{h-h'}|$, $|G_h|$, $|G_{h'}|$ and $|G_{h-h'}|$, their corresponding phases are ϕ_{-h}, $\phi_{h'}$, $\Phi_{h-h'}$, ψ_{-h}, $\psi_{h'}$ and $\psi_{h-h'}$. With

$$
|E_h| = R_1, \ |E_{h'}| = R_2, \ |E_{h-h'}| = R_3, \ |G_h| = S_1, \ |G_{h'}| = S_2, \ |G_{h-h'}| = S_3
$$

and

$$
\phi_{-h} = \phi_1, \ \phi_h = \phi_2, \ \Phi_{h-h'} = \phi_3, \ \psi_{-h} = \psi_1, \ \psi_{h'} = \psi_2, \ \psi_{h-h'} = \psi_3
$$

the joint probability distribution of the six normalised structure factors E_{-h}, $E_{h'}$, $E_{h-h'}$, G_{-h}, $G_{h'}$ and $G_{h-h'}$ is found to be

$$
\begin{aligned}
P(R_1, R_2, R_3, & S_1, S_2, S_3; \phi_1, \phi_2, \phi_3, \psi_1, \psi_2, \psi_3) = \\
& [1/\pi^6(1-\alpha^2)] R_1 R_2 R_3 S_1 S_2 S_3 \\
& \times \exp\{ -[1/(1-\alpha^2)](R_1^2 + R_2^2 + R_3^2 + S_1^2 + S_2^2 + S_3^2) \\
& + 2\beta[R_1 S_1 \cos(\phi_1 - \psi_1) + R_2 S_2 \cos(\phi_2 - \psi_2) + R_3 S_3 \cos(\phi_3 - \psi_3)] \\
& + 2\beta_0 R_1 R_2 R_3 \cos(\phi_1 + \phi_2 + \phi_3) + 2\beta_1[R_1 R_2 S_3 \cos(\phi_1 + \phi_2 + \psi_3) \\
& + R_1 S_2 R_3 \cos(\phi_1 + \psi_2 + \phi_3) + S_1 R_2 R_3 \cos(\psi_1 + \phi_2 + \phi_3)] \\
& + 2\beta_2[R_1 S_2 S_3 \cos(\phi_1 + \psi_2 + \psi_3) + S_1 R_2 S_3 \cos(\psi_1 + \phi_2 + \psi_3) \\
& + S_1 S_2 R_3 \cos(\psi_1 + \psi_2 + \phi_3)] + 2\beta_3 S_1 S_2 S_3 \cos(\psi_1 + \psi_2 + \psi_3)\},
\end{aligned}
\tag{5.23}
$$

where

$$
\alpha = \frac{\alpha_{11}}{\alpha_{20}^{1/2} \alpha_{02}^{1/2}},
$$

$$
\beta = \alpha/(1-\alpha^2),
$$

$$
\beta_0 = \frac{1}{\alpha_{20}^{3/2} \alpha_{02}^3 (1-\alpha^2)^3} (\alpha_{30}\alpha_{02}^3 - 3\alpha_{21}\alpha_{02}^2\alpha_{11} + 3\alpha_{12}\alpha_{02}\alpha_{11}^2 - \alpha_{03}\alpha_{11}^3),
$$

$$
\begin{aligned}
\beta_1 = \frac{1}{\alpha_{20}^2 \alpha_{02}^{5/2} (1-\alpha^2)^3} & [(\alpha_{21}\alpha_{20} - \alpha_{30}\alpha_{11})\alpha_{02}^2 - 2(\alpha_{12}\alpha_{20} - \alpha_{21}\alpha_{11})\alpha_{02}\alpha_{11} \\
& + (\alpha_{03}\alpha_{20} - \alpha_{12}\alpha_{11})\alpha_{11}^2],
\end{aligned}
$$

$$\beta_2 = \frac{1}{\alpha_{20}^{5/2}\alpha_{02}^2(1-\alpha^2)^3}\left[(\alpha_{12}\alpha_{02}-\alpha_{03}\alpha_{11})\alpha_{20}^2 - 2(\alpha_{21}\alpha_{02}-\alpha_{12}\alpha_{11})\alpha_{20}\alpha_{11}\right.$$
$$\left. + (\alpha_{30}\alpha_{02}-\alpha_{21}\alpha_{11})\alpha_{11}^2\right],$$

$$\beta_3 = \frac{1}{\alpha_{20}^3\alpha_{02}^{3/2}(1-\alpha^2)^3}(\alpha_{03}\alpha_{20}^3 - 3\alpha_{12}\alpha_{20}^2\alpha_{11} + 3\alpha_{21}\alpha_{20}\alpha_{11}^2 - \alpha_{30}\alpha_{11}^3). \quad (5.24)$$

The conditional probability distribution of the three-phase structure invariant $\omega_0 = \phi_{-h} + \phi_{h'} + \phi_{h-h'}$ given the values of $|E_{-h}|$, $|E_{h'}|$, $|E_{h-h'}|$, $|G_{-h}|$, $|G_{h'}|$ and $|G_{h-h'}|$ is denoted by

$$P(\Omega_0|R_1, R_2, R_3, S_1, S_2, S_3),$$

which is derived from (5.23) by fixing R_1, R_2, R_3, S_1, S_2 and S_3, integrating with respect to ψ_1, ψ_2 and ψ_3 from 0 to 2π and multiplying by a suitable normalising factor. The result is

$$P(\Omega_0|R_1, R_2, R_3, S_1, S_2, S_3) \simeq \frac{1}{K_0}\exp(A_0\cos\Omega_0). \quad (5.25)$$

Similarly we can obtain the conditional probability distribution for ω_1, ω_2 and ω_3. All the results can be expressed uniformly as

$$P(\Omega_i|R_1, R_2, R_3, S_1, S_2, S_3) \simeq \frac{1}{K_i}\exp(A_i\cos\Omega_i), \quad (5.26)$$

where

$$i = 0,1,2,3,$$
$$K_0 = 2\pi I_0(A_0), \quad (5.27)$$

$$A_0 = 2\{\beta_0 R_1 R_2 R_3 + \beta_1[R_1 R_2 R_3 T(2\beta R_3 S_3) + R_1 S_2 R_3 T(2\beta R_2 S_2)$$
$$+ S_1 R_2 R_3 T(2\beta R_1 S_1)] + \beta_2[R_1 S_2 S_3 T(2\beta R_2 S_2)T(2\beta R_3 S_3)$$
$$+ S_1 R_2 S_3 T(2\beta R_1 S_1)T(2\beta R_3 S_3) + S_1 S_2 R_3 T(2\beta R_1 S_1)T(2\beta R_2 S_2)]$$
$$+ \beta_3 S_1 S_2 S_3 T(2\beta R_1 S_1)T(2\beta R_2 S_2)T(2\beta R_3 S_3)\}, \quad (5.28)$$

$$K_1 = 2\pi I_0(A_1), \quad (5.29)$$

$$A_1 = 2\{\beta_1 R_1 R_2 S_3 + \beta_0 R_1 R_2 R_3 T(2\beta R_3 S_3) + \beta_2[R_1 S_2 S_3 T(2\beta R_2 S_2)$$
$$+ S_1 R_2 S_3 T(2\beta R_1 S_1)] + \beta_1[R_1 S_2 R_3 T(2\beta R_2 S_2)T(2\beta R_3 S_3)$$
$$+ S_1 R_2 R_3 T(2\beta R_1 S_1)T(2\beta R_3 S_3)] + \beta_3 S_1 S_2 S_3 T(2\beta R_1 S_1)T(2\beta R_2 S_2)$$
$$+ \beta_2 S_1 S_2 R_3 T(2\beta R_1 S_1)T(2\beta R_2 S_2)T(2\beta R_3 S_3)\}, \quad (5.30)$$

$$K_2 = 2\pi I_0(A_2), \quad (5.31)$$

$$A_2 = 2\{\beta_2 R_1 S_2 S_3 + \beta_1 [R_1 R_2 S_3 T(2\beta R_2 S_2) + R_1 S_2 R_3 T(2\beta R_3 S_3)]$$
$$+ \beta_3 S_1 S_2 S_3 T(2\beta R_1 S_1)] + \beta_0 R_1 R_2 R_3 T(2\beta R_2 S_2) T(2\beta R_3 S_3)$$
$$+ \beta_2 [S_1 R_2 S_3 T(2\beta R_1 S_1) T(2\beta R_2 S_2) + S_1 S_2 R_3 T(2\beta R_1 S_1) T(2\beta R_3 S_3)]$$
$$+ \beta_1 S_1 R_2 R_3 T(2\beta R_1 S_1) T(2\beta R_2 S_2) T(2\beta R_3 S_3)\}, \tag{5.32}$$

$$K_3 = 2\pi I_0(A_3), \tag{5.33}$$

$$A_3 = 2\{\beta_3 S_1 S_2 S_3 + \beta_2 [R_1 S_2 S_3 T(2\beta R_1 S_1) + S_1 R_2 S_3 T(2\beta R_2 S_2)$$
$$+ S_1 S_2 R_3 T(2\beta R_3 S_3)] + \beta_1 [R_1 R_2 S_3 T(2\beta R_1 S_1) T(2\beta R_2 S_2)$$
$$+ R_1 S_2 R_3 T(2\beta R_1 S_1) T(2\beta R_3 S_3) + S_1 R_2 R_3 T(2\beta R_2 S_2) T(2\beta R_3 S_3)]$$
$$+ \beta_0 R_1 R_2 R_3 T(2\beta R_1 S_1) T(2\beta R_2 S_2) T(2\beta R_3 S_3)\}. \tag{5.34}$$

In formulae (5.28), (5.30), (5.32) and (5.34), the function T is the ratio of two modified Bessel functions

$$T(z) = \frac{I_1(z)}{I_0(z)}. \tag{5.35}$$

It is obvious that $P(\Omega_i | R_1, R_2, R_3, S_1, S_2, S_3)$ has a unique maximum at $\Omega_i = 0$ or π according to $A_i > 0$ or $A_i < 0$ respectively. In addition the larger the magnitudes of A_i, the sharper is the distribution. Test calculations (Hauptman, Potter and Weeks, 1982) with the error-free data of the protein cytochrome C_{550} and a heavy-atom derivative showed that the formulae are capable of reliably identifying a large number of three-phase structure invariants having values close to either 0 or π. On the other hand, a test by Xu, Yang, Furey, Sax, Rose and Wang (1984) with error-free SIR data of the Bence Jones protein Rhe revealed that Hauptman's formulae tend to produce phases close to the averaged SIR phase – as was the case with Karle's algebraic formula (§5.3.1). The reason may be that, in the derivation of Hauptman's formulae, one has to use the probability distribution of the two-phase structure invariant $\phi_h - \psi_h$, the most probable value of which is zero as derived by Hauptman (1982*a*). This implies that most of the resulting phases should have a value close to either $\phi_R(\mathbf{h})$, the phase from the isomorphously replaced atoms, or $\phi_R(\mathbf{h}) + \pi$. Improvements of Hauptman's formulae were made by several authors (Fortier, Moore and Fraser, 1985; Klop, Krabbendam and Kroon, 1987; Hao and Fan, 1988). All of them used information from the known positions of the isomorphously replaced atoms rather than the expected value of the two-phase invariant $\phi_h - \psi_h$. Test results on this kind of method with experimental protein diffraction data have yet to be reported.

5.3.3 Combining direct methods with macromolecular methods

There have been two different approaches in applying direct methods to the phasing of protein SIR data. The first is to use direct methods alone while the second is to use a combination of direct methods with existing macromolecular methods. Procedures described in the previous section belong to the first category while those to be given in this section belong to the second. The philosophy of the latter procedures is to start direct methods from the end of traditional macromolecular methods – in other words, direct methods are used to derive not the phase $\phi(\mathbf{h})$ but rather the sign of $\Delta\phi(\mathbf{h})$ in (5.3).

According to Fan and Gu (1985) the probability distribution of $\Delta\phi(\mathbf{h})$ can be obtained by the product of Cochran and Sim distributions. We write the Cochran distribution (Cochran, 1955)

$$P_{\text{Cochran}}(\phi_h) = C\exp\left(\sum_{h'} K_{h,h'}\cos(\phi_h - \phi_{h'} - \phi_{h-h'})\right), \qquad (5.36)$$

where C is a normalising factor,

$$K_{h,h'} = 2\sigma_3\sigma_2^{-3/2}|E_hE_{h'}E_{h-h'}|,$$

$$\sigma_n = \sum_{j=1}^{N} z_j^n$$

and z_j is the atomic number of the jth atom in the unit cell. Let

$$\phi_h = \phi_{h,\text{R}} + \Delta\phi_h, \qquad (5.37)$$

where ϕ_h denotes the phase of structure factors from the native protein, $\phi_{h,\text{R}}$ is the phase contributed from the isomorphously replaced atoms and $\Delta\phi_h$ is the phase difference between ϕ_h and $\phi_{h,\text{R}}$. The absolute value of $\Delta\phi_h$ can be calculated as

$$|\Delta\phi_h| = \cos^{-1}\left[(|F_{h,\text{D}}|^2 - |F_{h,\text{R}}|^2 - |F_{h,\text{N}}|^2)/2|F_{h,\text{R}}F_{h,\text{N}}|\right] \qquad (5.38)$$

where $F_{h,\text{D}}$, $F_{h,\text{R}}$ and $F_{h,\text{N}}$ are the structure factors of the heavy-atom derivative, the partial structure of the replacing atoms and the native protein respectively; these quantities form a triangle in the complex plane. From (5.37) we have

$$\phi_h - \phi_{h'} - \phi_{h-h'} = \Delta\phi_h - (\Phi_3' + \Delta\phi_{h'} + \Delta\phi_{h-h'}) \equiv \Delta\phi_h - \beta'', \qquad (5.39)$$

where

$$\Phi_3' = -\phi_{h,\text{R}} + \phi_{h',\text{R}} + \phi_{h-h',\text{R}}.$$

Substitution of (5.39) into (5.36) gives

$$P_{\text{Cochran}}(\Delta\phi_h) = C \exp\left(\sum_{h'} K_{h,h'} \cos(\Delta\phi_h - \beta'')\right). \qquad (5.40)$$

Writing

$$\alpha' \sin\beta' = \sum_{h'} K_{h,h'} \sin\beta'',$$

$$\alpha' \cos\beta' = \sum_{h'} K_{h,h'} \cos\beta'',$$

(5.40) becomes

$$P_{\text{Cochran}}(\Delta\phi_h) = [2\pi I_0(\alpha')]^{-1} \exp[\alpha' \cos(\Delta\phi_h - \beta')], \qquad (5.41)$$

where $I_0(\alpha')$ is the zeroth-order modified Bessel function of the first kind with α' as argument. We also find

$$\alpha' = \left[\left(\sum_{h'} K_{h,h'} \sin\beta''\right)^2 + \left(\sum_{h} K_{h,h'} \cos\beta''\right)^2\right]^{1/2},$$

$$\tan\beta' = \sum_{h'} K_{h,h'} \sin\beta'' / \sum_{h'} K_{h,h'} \cos\beta''.$$

On the other hand, according to Sim (1960), if the partial structure of the isomorphously replaced atoms is known, we have

$$P_{\text{Sim}}(\Delta\phi_h) = [2\pi I_0(\kappa)]^{-1} \exp[\kappa \cos(\Delta\phi_h)], \qquad (5.42)$$

where

$$\kappa = 2|E_h E_{h,\text{R}}|/\sigma_u, \qquad \sigma_u = \sum_u z_u^2/\sigma_2,$$

$E_{h,\text{R}}$ is the contribution of the replacing atoms to the normalised structure factor and z_u is the atomic number of the uth atom belonging to the unknown part of the structure. Combination of (5.41) and (5.42) gives the total probability distribution of $\Delta\phi_h$, properly normalised, as

$$P(\Delta\phi_h) = P_{\text{Cochran}} \cdot P_{\text{Sim}} = [2\pi I_0(\alpha)]^{-1} \exp[\alpha \cos(\Delta\phi_h - \beta)], \qquad (5.43)$$

where

$$\alpha = \left[\left(\sum_{h'} K_{h,h'} \sin\beta''\right)^2 + \left(\sum_{h'} K_{h,h'} \cos\beta'' + \kappa\right)^2\right]^{1/2},$$

$$\tan\beta = \sum_{h'} K_{h,h'} \sin\beta'' \Big/ \left(\sum_{h'} K_{h,h'} \cos\beta'' + \kappa\right).$$

According to (5.38) $|\Delta\phi_h|$ is a known quantity when phase doublet information is available. Hence the probability that $\Delta\phi_h$ is positive can be derived from (5.43)

$$P_+(\Delta\phi_h) = \tfrac{1}{2} + \tfrac{1}{2}\tanh\left\{\sin|\Delta\phi_h|\left[\sum_{h'}K_{h,h'}\sin(\Phi'_3 + \Delta\phi_{h'} + \Delta\phi_{h-h'})\right]\right\}. \quad (5.44)$$

On the other hand, by maximising (5.43) we have $\Delta\phi_h = \beta$. Hence

$$\tan(\Delta\phi_h) = \frac{\displaystyle\sum_{h'}K_{h,h'}\sin(\Phi'_3 + \Delta\phi_{h'} + \Delta\phi_{h-h'})}{\displaystyle\sum_{h'}K_{h,h'}\cos(\Phi'_3 + \Delta\phi_{h'} + \Delta\phi_{h-h'}) + \kappa}. \quad (5.45)$$

The sign ambiguity is thus resolved in principle by (5.44) and (5.45).

In dealing with experimental data, it is important properly to take account of the experimental errors. The way of treating errors in protein crystallography as developed by Blow and Crick (1959) has been carried over to direct methods (Fan, Han and Qian, 1984). The treatment used here is somewhat modified from that of Fan *et al.* but leads to the same result. The basic assumption is that we have available a measured $|E_h|$ and a probability distribution for the phase angle, $P(\alpha)$, as shown in fig. 4.6. The phase to be found, the *best* phase, is denoted by α_{best}, and the true, but unknown, phase as α_t. The magnitude squared of the error in the structure factor will clearly be

$$|\Delta E_h|^2 = |E_h|^2\left|e^{i\alpha_{best}} - e^{i\alpha_t}\right|^2$$
$$= 2|E_h|^2[1 - \cos(\alpha_{best} - \alpha_t)]. \quad (5.46)$$

Since we do not know α_t but only its probability distribution, we can only find an expectation value of the left-hand side of (5.46). This is

$$\langle|\Delta E_h|^2\rangle = 2|E_h|^2\int_0^{2\pi}[1 - \cos(\alpha_{best} - \alpha)]P(\alpha)d\alpha. \quad (5.47)$$

The angle α_{best} is the *best* angle in the sense that it will lead to a minimum value of $\langle|\Delta E_h|^2\rangle$, so applying the condition

$$\frac{\partial\langle|\Delta E_h|^2\rangle}{\partial\alpha_{best}} = 0$$

gives

$$\int_0^{2\pi}\sin(\alpha_{best} - \alpha)P(\alpha)d\alpha = 0. \quad (5.48)$$

Hence

$$\sin \alpha_{\text{best}} \int_0^{2\pi} \cos \alpha \, P(\alpha) \mathrm{d}\alpha = \cos \alpha_{\text{best}} \int_0^{2\pi} \sin \alpha \, P(\alpha) \mathrm{d}\alpha \qquad (5.49)$$

or

$$\tan \alpha_{\text{best}} = \frac{\int_0^{2\pi} \sin \alpha \, P(\alpha) \mathrm{d}\alpha}{\int_0^{2\pi} \cos \alpha \, P(\alpha) \mathrm{d}\alpha} = \frac{T}{B}. \qquad (5.50)$$

If $P(\alpha)$ were a delta function then $T^2 + B^2$ would equal 1 and the value of α_{best} would be that indicated by the delta function. At the other extreme, if $P(\alpha)$ were a constant in the range 0 to 2π then T and B would both equal zero and no value of α_{best} would be indicated. In general, if $T^2 + B^2 = m_h^2$, then (5.50) is equivalent to

$$m_h \exp(\mathrm{i}\alpha_{\text{best}}) = \int_0^{2\pi} \exp(\mathrm{i}\alpha) \, P(\alpha) \mathrm{d}\alpha. \qquad (5.51)$$

What we are given by (5.51) is not only a *best* phase but also a figure-of-merit, m_h, to be associated with it where $0 \le m_h \le 1$. These are similar to quantities given by Blow and Crick (1959); in order to produce a *best* Fourier synthesis the structure factor of index $\mathbf{h}$ to be used is

$$E_{h,\text{best}} = m_h |E_h| \exp(\mathrm{i}\alpha_{h,\text{best}}). \qquad (5.52)$$

The probability distribution of the phase α_h corresponding to a phase doublet in the SIR case may be approximately simulated by the sum of two Gaussian functions with their maxima at $\alpha_h = \phi_1 = \phi_{h,R} + |\Delta\phi_h|$ and $\alpha_h = \phi_2 = \phi_{h,R} - |\Delta\phi_h|$, respectively, so that

$$P(\alpha_h) = \frac{1}{2\sigma_h(2\pi)^{1/2}} \exp[-(\alpha_h - \phi_1)^2/2\sigma_h^2]$$
$$+ \frac{1}{2\sigma_h(2\pi)^{1/2}} \exp[-(\alpha_h - \phi_2)^2/2\sigma_h^2] \qquad (5.53)$$

where σ_h can be obtained from the standard deviation, E, of the *lack of closure error* (Blow and Crick, 1959; §4.1.5). The way that σ_h is found may be seen from fig. 4.5, taking account of the difference in nomenclature now being used so that $F_{h,D} \equiv F_{PH}(\mathbf{h})$, $F_{h,N} \equiv F_P(\mathbf{h})$ and $F_{h,R} \equiv F_H(\mathbf{h})$. Blow and Crick showed that it was acceptable to assign the *lack of closure error* completely to $|F_{h,D}|$, the structure amplitude measured for the derivative. This being so, we see that

$$(|F_{h,D}| + x)^2 = |F_{h,N}|^2 + |F_{h,R}|^2 - 2|F_{h,N}||F_{h,R}| \cos \alpha. \qquad (5.54)$$

The quantity x is approximated as having a normal distribution with zero mean and standard deviation, E (see (4.9)). When $x = 0$ then the error in α, dα, will also be zero and we can identify the angle α, from fig. 4.5, as $\pi - \Delta\phi_h$. Differentiating both sides of (5.54) with respect to x

$$2(|F_{h,D}| + x) = 2|F_{h,N}||F_{h,R}| \sin \alpha \frac{d\alpha}{dx}. \qquad (5.55)$$

Assuming that $|F_{h,D}| \gg x$ and that $\sin \alpha \simeq \sin \Delta\phi_h$ over the effective range of variation of x, this gives

$$\alpha = \Delta\phi_h + \frac{|F_{h,D}|}{|F_{h,N}||F_{h,R}| \sin \Delta\phi_h} x. \qquad (5.56)$$

It follows from this that

$$\sigma_h^2 = \frac{|F_{h,D}|^2 E^2}{|F_{h,N}|^2 |F_{h,R}|^2 \sin^2 \Delta\phi_h}. \qquad (5.57)$$

If, for some reason, the probability for $\Delta\phi_h$ to be positive, P_+, does not equal that for it to be negative, P_-, then (5.53) can be written as

$$P(\alpha_h) = \frac{P_+}{2\sigma_h (2\pi)^{1/2}} \exp[-(\alpha_h - \phi_1)^2 / 2\sigma_h^2]$$

$$+ \frac{P_-}{2\sigma_h (2\pi)^{1/2}} \exp[-(\alpha_h - \phi_2)^2 / 2\sigma_h^2]. \qquad (5.58)$$

According to (5.51), we have

$$m_h \sin \alpha_{h,best} = \int_{-\infty}^{\infty} \sin \alpha_h P(\alpha_h) \, d\alpha_h, \qquad (5.59)$$

$$m_h \cos \alpha_{h,best} = \int_{-\infty}^{\infty} \cos \alpha_h P(\alpha_h) \, d\alpha_h. \qquad (5.60)$$

Substituting (5.58) into (5.59) and (5.60) and using the result

$$\int_0^\infty \cos(bx)\exp(-a^2x^2)\,dx = \frac{\pi^{\frac{1}{2}}}{2a}\exp(-b^2/4a^2),$$

one obtains

$$m_h \sin \alpha_{h,\text{best}} = \exp(-\sigma_h^2/2)(P_+ \sin \phi_1 + P_- \sin \phi_2), \tag{5.61}$$

$$m_h \cos \alpha_{h,\text{best}} = \exp(-\sigma_h^2/2)(P_+ \cos \phi_1 + P_- \cos \phi_2). \tag{5.62}$$

Dividing (5.61) by (5.62), it follows that

$$\tan \alpha_{h,\text{best}} = \frac{P_+ \sin \phi_1 + P_- \sin \phi_2}{P_+ \cos \phi_1 + P_- \cos \phi_2}. \tag{5.63}$$

In dealing with SIR data, it is more convenient to handle the phase difference, $\Delta\phi_h$, than the phase, ϕ_h, itself. Defining

$$\Delta\phi_{h,\text{best}} = \alpha_{h,\text{best}} - \phi_{h,R},$$

(5.63) can be simplified to

$$\tan(\Delta\phi_{h,\text{best}}) = \frac{2(P_+ - \frac{1}{2})\sin|\Delta\phi_h|}{\cos\Delta\phi_h}. \tag{5.64}$$

Adding together the squares of (5.61) and (5.62), one finds

$$m_h = \exp(-\sigma_h^2/2)\left[P_+^2 + P_-^2 + 2P_+P_-\cos(2\Delta\phi_h)\right]^{\frac{1}{2}} \tag{5.65}$$

or equivalently

$$m_h = \exp(-\sigma_h^2/2)\{[2(P_+ - \frac{1}{2})^2 + \frac{1}{2}][1 - \cos(2\Delta\phi_h)] + \cos(2\Delta\phi_h)\}^{\frac{1}{2}}. \tag{5.66}$$

m_h may be regarded as a measure of reliability of $\Delta\phi_{h,\text{best}}$. There are three factors affecting m_h:

(i) $\exp(-\sigma_h^2/2)$, a measure of the sharpness of the distribution of α_h.
(ii) $(P_+ - \frac{1}{2})^2$, a measure of the bias of $\Delta\phi_h$ towards positive or negative. It reaches its maximum value when P_+ equals 0 or 1.
(iii) $\cos(2\Delta\phi_h)$, a measure of the closeness of the two possible phases ϕ_1 and ϕ_2. It reaches its maximum value when $\Delta\phi_h$ equals 0 or π.

Either of the last two factors will have no effect on m_h when the other one reaches its maximum value. If $P_+ = P_- = \frac{1}{2}$ then (5.63) reduces to

$$\alpha_{h,\text{best}} = \begin{cases} \phi_{h,R}, & \text{if SIGN}(\cos\Delta\phi_h) = 1 \\ \phi_{h,R} + \pi, & \text{if SIGN}(\cos\Delta\phi_h) = -1 \end{cases}$$

or

$$\exp(i\alpha_{h,\,best}) = SIGN(\cos\Delta\phi_h)\exp(i\phi_{h,R}), \tag{5.67}$$

where $SIGN(\cos\Delta\phi_h)$ means 'the sign of $\cos\Delta\phi_h$'.

Meanwhile, (5.66) reduces to

$$m_h = \exp(-\sigma_h^2/2)|\cos\Delta\phi_h|. \tag{5.68}$$

Substituting (5.67) and (5.68) into (5.52), one obtains

$$E_{h,\,best} = \exp(-\sigma_h^2/2)\cos\Delta\phi_h |E_h|\exp(i\phi_{h,R}). \tag{5.69}$$

This is the *best* normalised structure factor that could be obtained just from SIR data. Now replacing the E_h terms in (5.44) and (5.45) by their *best* values, we have finally

$$P_+(\Delta\phi_h) = \tfrac{1}{2} + \tfrac{1}{2}\tanh\left\{\sin|\Delta\phi_h| \times\left[\sum_{h'} m_{h'} m_{h-h'} K_{h,h'}\sin(\Phi_3' + \Delta\phi_{h',best} + \Delta\phi_{h-h',best})\right]\right\}, \tag{5.70}$$

$$\tan(\Delta\phi_h) = \frac{\displaystyle\sum_{h'} m_{h'} m_{h-h'} K_{h,h'}\sin(\Phi_3' + \Delta\phi_{h',best} + \Delta\phi_{h-h',best})}{\displaystyle\sum_{h'} m_{h'} m_{h-h'} K_{h,h'}\cos(\Phi_3' + \Delta\phi_{h',best} + \Delta\phi_{h-h',best}) + \kappa}. \tag{5.71}$$

An iterative procedure was proposed to use (5.70) and (5.71) to break the SIR phase ambiguity. The flow-chart of the procedure is shown in fig. 5.2.

The efficiency of this method is illustrated by the two examples given below. The first example shows the application to a set of error-free SIR data with the isomorphously replaced atoms in a centrosymmetric arrangement, while the second is for a set of experimental SIR data with the isomorphously replaced atoms in a non-centrosymmetric arrangement.

1. Example 1 (Yao and Fan, 1985). Error-free SIR data were calculated for the model structure of aPP (avian pancreatic polypeptide) and that of its Hg derivative (Blundell, Pitts, Tickle, Wood and Wu, 1981). The crystals belong to space group C2 with unit cell parameters $a = 34.18$ Å, $b = 32.92$ Å, $c = 28.44$ Å and $\beta = 105.30°$. The asymmetric unit of the derivative contains only one isomorphously replaced atom, leading to a centrosymmetric arrangement. In this case, the three-phase invariants Φ_3' $(= -\phi_{h,R} + \phi_{h',R} + \phi_{h-h',R})$ contributed by the isomorphously replaced atoms will always equal 0 or π. The $\Delta\phi_{h,\,best}$ calculated from (5.64) at the beginning of iteration will also equal 0 or π for all $\mathbf{h}$, since all reflections will then have $P_+ = \tfrac{1}{2}$ so that the value of $P_+(\Delta\phi_h)$ calculated from (5.70)

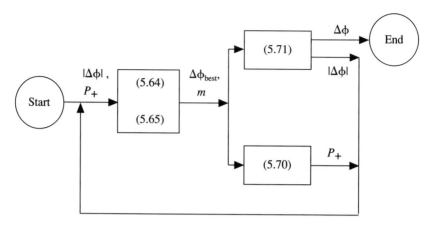

Fig. 5.2 Flow chart of the direct-method procedure for breaking the SIR phase ambiguity.

will always equal $\frac{1}{2}$. This will cause problems for Wang's method (see §5.2). In order to overcome this difficulty, a multi-solution procedure with random starting sign sets was used. Positive or negative signs associated with probability $P_+ = 0.6$ or 0.4 respectively were assigned randomly to the $|\varDelta\phi_h|$ terms. The signs were then refined by the iterative calculation using (5.64), (5.65) and (5.70). Of the 2100 independent reflections within 2 Å resolution, 1000 of the largest $|E|$ were used in the test. They yielded about 130 000 $\sum_2$ relationships, of which 60 000 were included in the calculation. The overall figure of merit

$$\mathrm{FOM} = \left(\sum_h m_h |E_h| / \sum_h \exp(-\sigma_h^2/2)|E_h \cos \varDelta\phi_h| \right) - 1 \qquad (5.72)$$

was used to predict the quality of the solutions. According to (5.66) and (5.72), if $P_+ = \frac{1}{2}$ for all E_h then we have FOM = 0 and a correct solution should have a large FOM value. Calculations were performed for ten random trials. The results are listed in table 5.1 in descending order of FOM. It will be seen that the three largest FOMs correspond to the best three resultant phase sets.

2. Example 2 (Fan and Wang, 1988, unpublished work). The experimental data used in this test were from the native Bence Jones protein Rhe and its Au-derivative (Wang, Yoo and Sax, 1979) at 3 Å resolution. The crystals belong to space group $P2_12_12$ with unit cell parameters $a = 54.6$ Å, $b = 52.2$ Å and $c = 42.6$ Å. Some 2044 independent reflections and 1 118 671 $\sum_2$ relationships were involved in the calculation. Results from

Table 5.1. *Test results with the error-free SIR data from the native protein aPP and its Hg-derivative. The reflections were arranged in descending order of* $|P_+ - \frac{1}{2}|$ *and then cumulated into four groups. The groups numbered 1, 2, 3 and 4 contain the top 200, 400, 600 and 800 reflections respectively. The percentage of reflections with the signs of* $\Delta\phi_h$ *correctly determined (%) and average error of phases (in degrees) are shown.*

| | | Reflection group | | | | | | | |
| | | 1 | | 2 | | 3 | | 4 | |
FOM	Trial set	%	Error	%	Error	%	Error	%	Error
0.4136	1	92.0	7	86.3	15	78.5	20	72.4	22
0.4091	8	90.5	9	83.3	18	78.0	20	72.8	21
0.3995	9	95.0	6	88.5	11	79.8	18	74.3	20
0.3918	2	89.5	10	81.8	18	74.8	22	70.6	23
0.3848	3	87.0	16	82.3	19	74.8	23	70.6	25
0.3690	10	80.0	21	75.3	26	68.7	30	64.1	30
0.3575	5	68.5	37	63.3	42	62.5	39	60.8	37
0.3518	4	59.5	49	57.0	49	56.0	46	53.6	43
0.3446	7	69.0	39	63.5	40	62.7	38	58.9	37
0.3362	6	66.5	42	60.3	44	60.2	42	56.9	39

Table 5.2. *Test results with the experimental SIR data from the native Bence Jones Rhe and its Au-derivative. Reflections were arranged in descending order of* d-*spacing* $(\lambda/2\sin\theta)$ *and then cumulated into three groups with resolution limit of 8 Å, 5 Å, and 3 Å respectively.*

| Resolution limit (Å) | Number of reflections | Average phase error in degrees | |
		Direct-method phases	SIR-phases
8.0	134	38.4	42.9
5.0	508	43.3	49.4
3.0	2044	46.5	51.7

the iterative calculation using (5.64), (5.65) and (5.70) are listed in tables 5.2 and 5.3. It is evident that the SIR phases were considerably improved by the direct-method phase-refinement technique.

These two examples are indicative of what can be achieved with the SIR method. Although it is likely that better results can be obtained with MIR

Table 5.3. *Test results with the experimental SIR data from the native Bence Jones protein Rhe and its Au-derivative. Reflections were arranged in descending order of F(obs) and then cumulated into five groups with F(obs) ≥ 800, 500, 300, 100 and 50 respectively.*

F(obs)	Number of reflections	Average phase error in degrees	
		Direct-method phases	SIR-phases
800	504	15.1	28.2
500	414	33.7	44.5
300	1248	41.9	48.8
100	2035	46.5	51.7
50	2044	46.5	51.7

data since, in general, more data should lead to better results, a great deal is possible with good SIR data. All that is necessary is to get an initial phase set of sufficient quality to enable the phase extension and refinement techniques described in chapter 7 to be deployed and if the SIR approach can do this then it may not be worthwhile to try to solve the sometimes very difficult experimental problems of obtaining extra heavy-atom derivatives.

6

Use of anomalous scattering data

6.1 Introduction

The anomalous scattering effect from specific atoms can be used in protein crystallography to locate heavy-atom sites, to determine the absolute configuration of the protein molecule in the crystal and to phase individual reflections for solving the crystal structure. The present chapter concentrates on this last application.

For anomalous scattering, we have

$$F^+ = F° + F' + F'' \tag{6.1}$$

$$F^- * = F° + F' - F'', \tag{6.2}$$

where F^+ and F^- are structure factors of a Friedel pair, $F^- *$ is the complex conjugate of F^-, $F°$ is the structure factor with all atoms in the unit cell scattering normally and F' is the real-part correction for the anomalous scatterers, while F'' is the imaginary-part correction for the same set of anomalous scatterers. Combining (6.1) and (6.2) we obtain

$$F^+ - F^- * = 2F''. \tag{6.3}$$

The magnitudes of F^+ and $F^- *$ can be measured from experiment. They can then be used to determine the positions of the anomalous scatterers (§4.2.4), which, in turn, can be used to derive the vector F''. Now there are two possible ways to draw the triangle determined by (6.3), these ways being symmetrically arranged with respect to the vector $2F''$ (fig. 6.1). This leads to two possible solutions for the structure factor $F°$, indicated by a thick solid line for the supposed true one and by a dashed line for the false one.

The two possible solutions from one-wavelength anomalous scattering, shown in fig. 6.1, can also be expressed as a phase doublet. We define a median structure factor

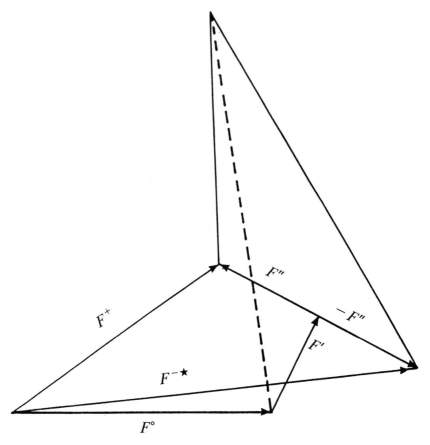

Fig. 6.1 Two possible solutions for the structure factor $F°$ resulting from one-wavelength anomalous scattering. The true solution is shown by a thick solid line while the false solution is denoted by a dashed line.

$$F = F° + F'. \tag{6.4}$$

The magnitude of F is given precisely by

$$|F|^2 = \tfrac{1}{2}(|F^+|^2 + |F^-|^2) - |F''|^2 \tag{6.5a}$$

or approximately by

$$|F| \simeq \tfrac{1}{2}(|F^+| + |F^-|) \tag{6.5b}$$

if the positions of the anomalous scatterers are not known.

Corresponding to the two solutions of $F°$ in fig. 6.1 there are two possible phases for F (see fig. 6.2), i.e.

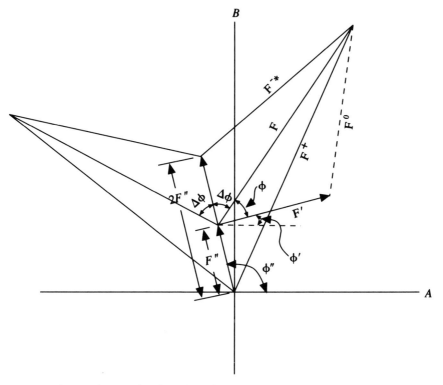

Fig. 6.2 Phase ambiguity arising from one-wavelength anomalous scattering.

$$\phi = \phi'' \pm |\varDelta\phi|, \tag{6.6}$$

where $|\varDelta\phi|$ can be calculated exactly by solving the triangle in fig. 6.2, giving

$$|\varDelta\phi| = \left| \cos^{-1} \left(\frac{|F^+|^2 - |F^-|^2}{4|F||F''|} \right) \right| \tag{6.7a}$$

or approximately as

$$|\varDelta\phi| \simeq \left| \cos^{-1} \left[(|F^+| - |F^-|)/2F'' \right] \right|. \tag{6.7b}$$

Equation (6.6) defines the phase ambiguity arising from one-wavelength anomalous scattering. The successful use of anomalous scattering in phase determination relies on resolving this ambiguity.

6.2 Combining anomalous scattering with isomorphous replacement

The complementary nature of anomalous scattering and isomorphous replacement is clearly seen by comparing fig. 6.2 with fig. 5.1. Suppose that the anomalous scatterers in the isomorphous derivative are themselves the isomorphously replaced atoms. Then F' in fig. 6.2 will be coincident with F_R in fig. 5.1. Suppose also, for simplicity, that there is only one kind of anomalous scatterer, which means that, in fig. 6.2, F' will be perpendicular to F''. Now the two possible solutions due to single isomorphous replacement will be symmetrically arranged with respect to F', while that of one-wavelength anomalous scattering will be symmetrically arranged with respect to F''. Among the total four possible solutions, the two true ones will be close together (coincident in an ideal case) while the others will, in general, be separated. Thus the combination of single isomorphous replacement and one-wavelength anomalous scattering leads to a unique solution of the phase problem. This was first proposed by Bijvoet (1949). A successful example of the use of this idea was provided by Blundell, Pitts, Tickle, Wood and Wu (1981) in solving the structure of the protein avian pancreatic polypeptide.

6.3 Multi-wavelength anomalous scattering

The phase ambiguity arising from the one-wavelength anomalous scattering can, in principle, be resolved by introducing a second wavelength. In the same way as for the first wavelength (see fig. 6.1), this also leads to two possible solutions. However, from the total of four possible solutions the two true solutions, one from each wavelength, should coincide and so be recognised (see fig. 6.3). It should be noticed that the two wavelengths must be chosen so that $F'_1 \neq F'_2$, otherwise F''_2 will coincide with F''_1 giving a coincidence of the two dashed lines (see fig. 6.3), thus leaving the ambiguity unresolved. From a practical point of view a larger F'' will give more accurate values for the possible solutions, while a larger difference of F' will provide better discrimination in resolving the ambiguity. The two-wavelength anomalous dispersion method was first proposed by Raman (1959b). The first wavelength was chosen to be slightly shorter than the absorption edge of the anomalous scatterers so that both f' and f'' are significant for the anomalous scatterers but negligible for the remaining atoms. The second wavelength was chosen far away from the absorption edges of all the atoms so that both f' and f'' are negligible for all the atoms. According to (6.1) we have for the first wavelength

$$|F^+|^2 = (F^\circ + F' + F'')(F^{\circ *} + F'^* + F''^*)$$
$$= |F^\circ|^2 + |F'|^2 + |F''|^2 + 2|F^\circ F'|\cos(\phi^\circ - \phi')$$
$$+ 2|F^\circ F''|\cos(\phi^\circ - \phi'') + 2|F'F''|\cos(\phi' - \phi''), \qquad (6.8)$$

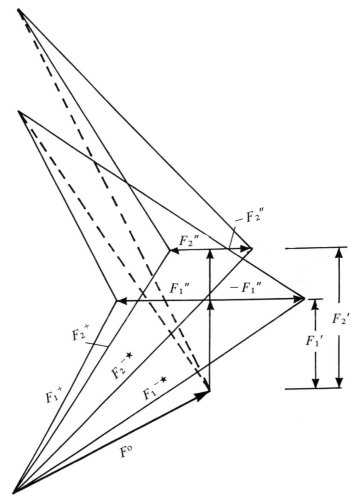

Fig. 6.3 Two-wavelength anomalous scattering. The two true solutions of $F°$, shown as the thick solid line, are coincident, while the two false solutions are separated, as shown by the dashed lines.

where $\phi°$, ϕ' and ϕ'' are the phases of $F°$, F' and F'' respectively. Similarly from (6.2) we have

$$|F^-|^2 = (F° + F' - F'')(F^{*°} + F^{*'} - F^{*''})$$
$$= |F°|^2 + |F'|^2 + |F''|^2 + 2|F°F'|\cos(\phi° - \phi')$$
$$- 2|F°F''|\cos(\phi° - \phi'') - 2|F'F''|\cos(\phi' - \phi''). \quad (6.9)$$

Combining (6.8) and (6.9) it follows that

$$|F^+|^2 + |F^-|^2 = 2|F^\circ|^2 + 2|F'|^2 + 2|F''|^2 + 4|F^\circ F'|\cos(\phi^\circ - \phi'), \quad (6.10)$$

$$|F^+|^2 - |F^-|^2 = 4|F^\circ F''|\cos(\phi^\circ - \phi'') + 4|F'F''|\cos(\phi' - \phi''). \quad (6.11)$$

If the positions of the anomalous scatterers are known, one can calculate the quantities $|F'|$, ϕ', $|F''|$ and ϕ'' for both wavelengths. In addition the quantities $|F^+|$ and $|F^-|$ can be measured from the experiment at the first wavelength while $|F^\circ|$ can be measured from that at the second wavelength so that the phase, ϕ°, of F° can be uniquely determined.

The advent of synchrotron radiation as a tunable source of X-rays has opened new prospects for the application of anomalous scattering. Data can be taken at wavelengths that optimise either f' or f'' and several different data sets can be collected. Methods have been developed to exploit the multiplicity of data sets and these are dealt with in detail in §6.5.

6.4 One-wavelength anomalous scattering

Great efforts have been made to resolve the ambiguity arising from one-wavelength anomalous scattering (OAS) without using additional diffraction data. This is of importance in protein crystallography since most protein crystals are sensitive to X-ray irradiation and isomorphous derivatives are not always easy to prepare.

6.4.1 Phase ambiguity resolved by using heavy-atom information

In contrast with single isomorphous replacement, the two possible solutions from one-wavelength anomalous scattering are not symmetrically arranged with respect to the heavy-atom contribution, which is parallel to F' (see figs. 6.1 and 6.2). Accordingly Ramachandran and Raman (1956) proposed that one can always make that choice which has a phase closer to that of F'. This is a combination of anomalous scattering and heavy-atom methods. Hendrickson and Teeter (1981) used a similar but improved method in the structure determination of the hydrophobic protein, crambin, by making use of anomalous scattering from sulphur. The molecule of crambin has three disulphide bridges among its 46 amino acid residues. The crystal structure was solved at 1.5 Å resolution directly from Cu $K\alpha$ ($\lambda = 1.54$ Å) diffraction data of a native crystal; the wavelength used was far away from the sulphur absorption edge (5.36 Å) so that anomalous scattering was far from being optimised. The crystals of crambin are in space group $P2_1$ with unit cell dimensions $a = 40.96$ Å, $b = 18.65$ Å, $c = 22.52$ Å and $\beta = 90.77°$. The sulphur positions were located by the interpretion of a Patterson map computed with coefficients $(|F^+| - |F^-|)^2$. The information from one-wavelength anomalous scattering is expressed as the bimodal probability distribution

$$P_A(\phi) = N \exp\{-[\Delta F - 2|F''|\sin(\phi - \phi')]^2 / 2\mathscr{E}^2\}. \tag{6.12}$$

Here, N is a normalisation factor; $\Delta F = |F^+| - |F^-|$ is the Bijvoet difference with expected value $2|F''|\sin(\phi - \phi')$, which can be deduced from fig. 6.2; and $\mathscr{E} = (\sigma_{\Delta F}^2 + \mathscr{E}_0^2)^{\frac{1}{2}}$ where $\mathscr{E}_0$ is the residual lack-of-closure error (Ten Eyck and Arnone, 1976).

On the other hand, the probability density based on the heavy atoms (sulphur), is expressed by the Sim distribution (Sim, 1960)

$$P_S(\phi) = N' \exp[2Q|F||F_S|\cos(\phi - \phi')/\langle F_U^2 \rangle]. \tag{6.13}$$

Here, N' is a normalization factor; $|F_S|$ is the structure factor magnitude calculated from the sulphur atoms; and $\langle F_U^2 \rangle$ is the expected square of the contribution from the unknown part of the structure (including all the atoms in the unit cell other than sulphur).

The phase for a particular reflection was determined as follows. If $P_S(\phi)$ of (6.13) clearly favoured one of the alternative maxima in $P_A(\phi)$ of (6.12) then the unimodal distribution, $P_S(\phi)$, was used directly. Otherwise, $P_A(\phi)$ and $P_S(\phi)$ were multiplicatively combined to give the distribution of the phase angle ϕ. Equations (6.12) and (6.13) were also used to calculate the figure of merit, m, as given in (5.51), for weighting the Fourier synthesis. Interpretation of the Fourier map led finally to the correct structural model.

6.4.2 The algebraic method

With a derivation similar to that for the SIR case (§5.3.1; Karle, 1983), simple rules can be obtained that permit the estimation of three-phase structure invariants in the OAS case (Karle, 1984). Assume that in the crystal structure there exists one type of predominant anomalous scatterer with atomic scattering factor

$$f = f^\circ + f' + f'', \tag{6.14}$$

where f° is the atomic scattering factor without the contribution from anomalous scattering and f' and f'' are respectively the real and the imaginary parts of the anomalous component of the atomic scattering factor. The rules obtained are then as follows.

$R_{\text{ano},1}$: if the sign of the product of the largest magnitude differences,

$$(|F_{-h}| - |F_h|)(|F_{h'}| - |F_{-h'}|)(|F_{h-h'}| - |F_{-h+h'}|),$$

is the same as the sign of f'', then the value of the average three-phase structure invariant is close to $-\pi/2$ and, when the signs are opposite, the value is close to $\pi/2$.

$R_{ano,2}$: if the sign of the product of the largest magnitude differences,

$$(|F_{-h}+F_h^*|-2|F_{-h}^\circ|)(|F_{h'}+F_{-h'}^*|-2|F_{h'}^\circ|)(|F_{h-h'}+F_{-h+h'}^*|-2|F_{h-h'}^\circ|),$$

is the same as the sign of f', then the value of the average three-phase structure invariant is close to zero and, when the signs are opposite, the value is close to π.

$R_{ano,3}$: if the sign of the product of the largest magnitude differences,

$$(|F_{-h}|-|F_{-h}^\circ|)(|F_{h'}|-|F_{h'}^\circ|)(|F_{h-h'}|-|F_{h-h'}^\circ|),$$

is positive, then the value of the average three-phase structure invariant is close to 3δ and when the sign is negative, the value is close to $3\delta+\pi$, where $\delta=\tan^{-1}(f''/f')$.

The value of $|F_h^\circ|$ in rule 2 and rule 3 is estimated from

$$|F_h^\circ|=\tfrac{1}{2}W(|F_h|+|F_{-h}|), \tag{6.15}$$

where

$$W=\left(\frac{\displaystyle\sum_{j=1}^{N_{ano}}f_j^2+\sum_{j=1}^{N_{ano}}(f_j^\circ)^2}{\displaystyle\sum_{j=1}^{N_{ano}}f_j^2+\sum_{j=1}^{N_{ano}}[(f_j^\circ+f')^2+f''^2]}\right)^{1/2} \tag{6.16}$$

and N_{ano} is the number of anomalous scatterers. With slight modification the rules can also be used in cases of more than one kind of anomalous scatterer. Test calculations showed that these rules have the potential for yielding reliable estimates of large numbers of three-phase structure invariants but no results of tests with experimental protein data have been reported.

6.4.3 Integration of direct methods with OAS data

Probability formulae integrating direct methods with OAS data were derived by Hauptman (1982b) and later by Giacovazzo (1983). Hauptman defines the normalised structure factor in the presence of anomalous scatterers as

$$E_h=\frac{1}{\alpha_h^{1/2}}\sum_{j=1}^{N'}f_{jh}\exp(i2\pi\mathbf{h}\cdot\mathbf{r}_j)$$

$$= \frac{1}{\alpha_h^{1/2}} \sum_{j=1}^{N} |f_{jh}| \exp[i(\delta_{jh} + 2\pi \mathbf{h} \cdot \mathbf{r}_j)], \tag{6.17}$$

where $f_{jh} = |f_{jh}| \exp(i\delta_{jh})$ and $\alpha_h = \sum_{j=1}^{N} |f_{jh}|^2$.

Let $-\mathbf{h}, \mathbf{h}'$ and $\mathbf{h} - \mathbf{h}'$ be three reciprocal-lattice vectors. This gives rise to eight distinct three-phase structure invariants:

$$\begin{aligned}
\Psi_0 &= \phi_{-h} + \phi_{h'} + \phi_{h-h'}, \\
\Psi_1 &= -\phi_h + \phi_{h'} + \phi_{h-h'}, \\
\Psi_2 &= \phi_{-h} - \phi_{-h'} + \phi_{h-h'}, \\
\Psi_3 &= \phi_{-h} + \phi_{h'} - \phi_{-h+h'}, \\
\Psi_{\bar{0}} &= \phi_h + \phi_{-h'} + \phi_{-h+h'}, \\
\Psi_{\bar{1}} &= -\phi_{-h} + \phi_{-h'} + \phi_{-h+h'}, \\
\Psi_{\bar{2}} &= \phi_h - \phi_{h'} + \phi_{-h+h'}, \\
\Psi_{\bar{3}} &= \phi_h + \phi_{-h'} - \phi_{h-h'}.
\end{aligned} \tag{6.18}$$

The first neighbourhood of each of the three-phase structure invariants is defined to consist of the six magnitudes $|E_{-h}|$, $|E_{h'}|$, $|E_{h-h'}|$, $|E_h|$, $|E_{-h'}|$, $|E_{-h+h'}|$. The reciprocal-lattice vectors $-\mathbf{h}, \mathbf{h}'$ and $\mathbf{h} - \mathbf{h}'$ are fixed and six non-negative numbers R_1, R_2, R_3, $R_{\bar{1}}$, $R_{\bar{2}}$ and $R_{\bar{3}}$ are specified. The N-fold Cartesian product W is defined as consisting of all ordered N-tuples $(\mathbf{r}_1, \mathbf{r}_2, \ldots, \mathbf{r}_N)$, where $\mathbf{r}_1, \mathbf{r}_2, \ldots, \mathbf{r}_N$ are atomic position vectors. Suppose that the primitive random variable is the N-tuple $(\mathbf{r}_1, \mathbf{r}_2, \ldots, \mathbf{r}_N)$, assumed to be uniformly distributed over the subset of W defined by

$$\begin{aligned}
|E_{-h}| &= R_1, |E_{h'}| = R_2, |E_{h-h'}| = R_3, \\
|E_h| &= R_{\bar{1}}, |E_{-h'}| = R_{\bar{2}}, |E_{-h+h'}| = R_{\bar{3}},
\end{aligned} \tag{6.19}$$

where the normalised structure factors E are defined by (6.17). Then the eight structure invariants Ψ_j defined in (6.18), as functions of primitive random variables $(\mathbf{r}_1, \mathbf{r}_2, \ldots, \mathbf{r}_N)$, are themselves random variables. Now denote by

$$P_j(\Omega_j | R_1, R_2, R_3, R_{\bar{1}}, R_{\bar{2}}, R_{\bar{3}}) = P_j(\Omega_j), \tag{6.20}$$
$$j = 0, 1, 2, 3, \bar{0}, \bar{1}, \bar{2}, \bar{3}$$

the conditional probability distribution of each Ψ_j in (6.18), assuming as known the six magnitudes (6.19) in its first neighbourhood. Then it can be shown that

$$P_j(\Omega_j) = \frac{1}{K_j} \exp[A_j \cos(\Omega_j - \omega_j), \tag{6.21}$$

$$j=0,1,2,3,\bar{0},\bar{1},\bar{2},\bar{3}.$$

K_j, A_j and ω_j can be calculated from the complex atomic scattering factors and the observed magnitudes $|E_{-h}|$, $|E_{h'}|$, $|E_{h-h'}|$, $|E_h|$, $|E_{-h'}|$ and $|E_{-h+h'}|$ (see §3.2 of Hauptman's original paper). No prior knowledge of the positions of the anomalous scatterers is needed, nor is it required that the anomalous scatterers be identical. Since the K_j and A_j terms are positive, the maximum of (6.21) occurs at $\Omega_j=\omega_j$. Hence, when the variance of the distribution (6.21) is small, i.e. A_j is large, there is obtained the reliable estimate

$$\Psi_j=\omega_j, \qquad j=0,1,2,3,\bar{0},\bar{1},\bar{2},\bar{3}. \tag{6.22}$$

With error-free protein data this method has been found to be capable of yielding unique estimates for a large number of three-phase structure invariants. No practical applications have been reported.

6.4.4 Wang's method

Wang's seven-step procedure described in §5.2 can also be applied in resolving the phase ambiguity due to OAS. However, modifications must be made to steps 1, 5 and 6. In step 1, phases and figures of merit for acentric reflections are calculated with the bimodal OAS phase probability distribution (Blow and Crick, 1959; Hendrickson and Lattman, 1970), while phases and figures of merit for centric reflections are initially set to zero and later assigned the values calculated from step 5. In step 6, the original bimodal OAS phase probability distribution, instead of the SIR phase distribution, is multiplied by the Sim distribution centred on the phases found in step 5. One of the differences between Wang's method and that of Hendrickson and Teeter (1981) is that Wang breaks the phase ambiguity by a rough structure model obtained from step 4, while Hendrickson and Teeter make use only of the partial structure of anomalous scatterers (§6.4.1).

Examples of practical application can be found with a bovine neurophysin II dipeptide complex (Chen, Rose, Breslow, Yang, Chang, Furey, Sax and Wang, 1991) and in the crystal structure analysis of Cd, Zn metallothionein (Robbins, McRee, Williamson, Collett, Xuong, Furey, Wang and Stout, 1991). In the first example, crystals are in space group $P2_12_12_1$ with $a=120.0$ Å, $b=69.4$ Å and $c=62.4$ Å. Diffraction data to 2.8 Å resolution were collected with Cu $K\alpha$ radiation and the positions of the anomalous scatterers, iodine, were located by interpreting the anomalous difference Patterson map calculated at 5 Å resolution. The phase ambiguity was resolved at 3.5 Å resolution while the complete structure was obtained at 2.8 Å resolution by Wang's procedure. In the second example, crystals are in space group $P4_32_12$ with unit cell dimensions $a=b=30.9$ Å and $c=120.4$ Å. Diffraction data to 2.0 Å resolution were collected with Cu Kα radiation. The five Cd positions were found by the direct-method programme MULTAN (§3.5.3) using $|F|$ instead of $|\Delta F|$ as input. Based on the refined Cd sites Wang's procedure resolved the OAS phase ambiguity and finally solved the structure.

6.4.5 *Phase ambiguity resolved by direct methods*

There were several early proposals to use direct methods to break the phase ambiguity inherent in the OAS technique (Fan, 1965b; Hazell, 1970; Sikka, 1973; Heinerman, Krabbendam, Kroon and Spek, 1978). The method of Fan (1965b) has been extended and tested with experimental protein diffraction data. Details of the methods are given here. The formulation in §5.3.3 can easily be made common for both the SIR and OAS cases (Fan and Gu, 1985). The phase doublet, from either SIR or OAS, is expressed as

$$\phi_h = \langle \phi_h \rangle \pm |\Delta\phi_h|, \tag{6.23}$$

where $\langle \phi_h \rangle$ is equal to $\phi_{h,R}$ or ϕ'', the phase of F'', while $|\Delta\phi_h|$ is calculated by (5.38) or (6.7) for SIR or OAS, respectively (see figs. 5.1 and 6.2). By introducing the concept of *best* phase, $\alpha_{h,best}$, and figure of merit, m_h, used in protein crystallography into the direct-method approach for dealing with enantiomorphous phase ambiguity (Fan, Han and Qian, 1984), there is obtained

$$\Delta\phi_{h,best} = \alpha_{h,best} - \langle \phi_h \rangle, \tag{6.24}$$

$$\tan(\Delta\phi_{h,best}) = \frac{2(P_+ - \tfrac{1}{2})\sin|\Delta\phi_h|}{\cos\Delta\phi_h}, \tag{6.25}$$

$$m_h = \exp(-\sigma_h^2/2)\{[2(P_+ - \tfrac{1}{2})^2 + \tfrac{1}{2}][1 - \cos(2\Delta\phi_h)] + \cos(2\Delta\phi_h)\}^{\frac{1}{2}}, \tag{6.26}$$

where σ_h is related to the experimental error and can be calculated from the mean square of the 'lack of closure error' (equation (5.57); Blow and Crick, 1959). The probability that $\Delta\phi_h$ is positive, P_+, is given by

$$P_+ = \tfrac{1}{2} + \tfrac{1}{2}\tanh\left\{\sin|\Delta\phi_h| \right.$$
$$\left. \times\left[\sum_{h'} m_h \cdot m_{h-h'} K_{h,h'} \sin(\Phi_3' + \Delta\phi_{h'best} + \Delta\phi_{h-h'best}) + \kappa\sin\delta_h\right]\right\}, \tag{6.27}$$

where, for the SIR or OAS cases respectively, we have

$$\Phi'_3 = -\phi_{h,\text{R}} + \phi_{h',\text{R}} + \phi_{h-h',\text{R}} \text{ or } -\phi''_h + \phi''_{h'} + \phi''_{h-h'},$$

$$\kappa = 2|E_h E_{h,\text{R}}|/\sigma_u \text{ or } 2|E_h E_{h,\text{A}}|/\sigma_u,$$

$$\delta_h = 0 \text{ or } \phi'_h - \phi''_h.$$

In the above expressions, $E_{h,\text{R}}$ is the contribution of the isomorphously replaced atoms while $E_{h,\text{A}}$ is the contribution of the anomalous scatterers to the normalised structure factor, $\sigma_u = \sum_u z_u^2/\sigma_2$, z_u is the atomic number of the uth atom belonging to the unknown part of the structure, $\sigma_2 = \sum_{j=1}^{N} z_j^2$ and N is the number of atoms in the unit cell.

A procedure for using (6.24)–(6.27) is now described. Values of $\Delta\phi_{h,\text{best}}$ and m_h are calculated for each reflection using (6.25) and (6.26), assuming that $P^+ = \frac{1}{2}$. The values of $\Delta\phi_{h,\text{best}}$ and m_h are then substituted into (6.27) to obtain for each reflection a new P_+, which will mostly differ from $\frac{1}{2}$. Substituting the new values of P_+ into (6.25) and (6.26) gives an improved set of $\Delta\phi_{h,\text{best}}$ and m_h. Next $\alpha_{h,\text{best}}$ is calculated from $\Delta\phi_{h,\text{best}}$ from (6.24) and values of $\alpha_{h,\text{best}}$ and m_h are then used with the observed structure-factor magnitudes to calculate the *best* Fourier map.

The above procedure has been tested (Fan, Hao, Gu, Qian, Zheng and Ke, 1990) with the experimental OAS data from the Hg-derivative of the protein aPP (Blundell, Pitts, Tickle, Wood and Wu, 1981). The sample crystallises in space group C2 with unit cell dimensions $a = 34.18$ Å, $b = 32.92$ Å, $c = 28.44$ Å and $\beta = 105.30°$ and with one molecule of 36 amino-acid residues in the asymmetric unit. Diffraction data were collected with Cu Kα radiation and 2108 independent reflections at 2 Å resolution were observed and used in the test calculation. The resultant direct-method phases led to an interpretable electron density map, a part of which is shown in fig. 6.4. The correlation coefficient between the electron density map phased by the direct method and that calculated from the true phases (§4.3) is 0.70. The mean phase error of the direct-method phases in comparison with the true phases is 38.4° for the total of 2108 independent reflections at 2 Å resolution.

Three different kinds of distributions are involved in the method described here. They are the bimodal phase probability distribution from the OAS data, the Sim distribution calculated with the partial structure of the anomalous scatterers and the Cochran distribution of three-phase structure invariants. The advantage of combining these three distributions is shown in fig. 6.5, in which the same part of the electron density map was

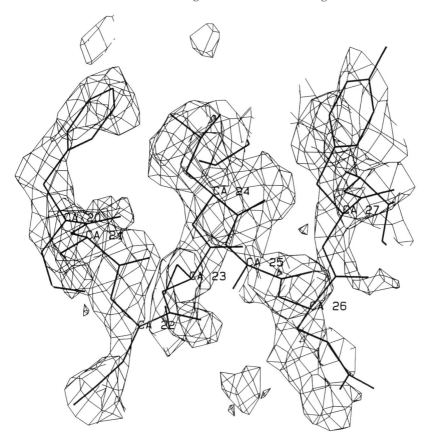

Fig. 6.4 Portion of the electron-density map for aPP calculated with phases derived from the use of equations (6.24)–(6.27).

calculated with the three different phasing methods: (a) phasing by the bimodal OAS distribution only; (b) phasing by combining the OAS distribution with the Sim distribution; and (c) phasing by combining the OAS, the Sim and the Cochran distribution. It can clearly be seen that phasing method (c) gives the best result.

6.4.6 *Phase ambiguity resolved by Wilson statistics*

Ralph and Woolfson (1991) proposed a method making use of Wilson statistics (Wilson, 1949). Assuming that the positions of the anomalous

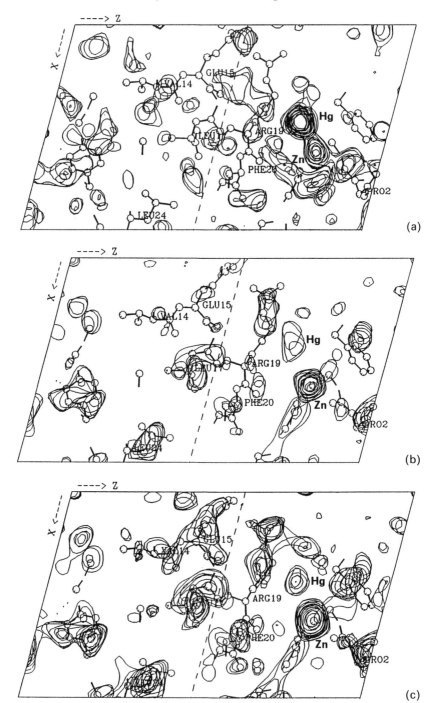

scatterers are known, the OAS ambiguity can be expressed in terms of the two possible contributions of the unknown part of the structure, i.e. F_{L1} and F_{L2} in fig. 6.6. Wilson statistics can then be applied to the unknown part of the structure to give the relative probabilities of $|F_{L1}|$ and $|F_{L2}|$. The required Wilson distributions are

$$P_{\bar{1}}(|F|) = (1/2\pi\Sigma)^{\frac{1}{2}}\exp(-|F|^2/2\Sigma) \qquad (6.28)$$

for a centric reflection and

$$P_1(|F|) = (2/\Sigma)|F|\exp(-|F|^2/\Sigma) \qquad (6.29)$$

for an acentric reflection where $\Sigma = \sum_{j=1}^N f_j^2$. Therefore the probability for $\Delta\phi$ to be positive is

$$P(+\Delta\phi) = P(|F_{L1}|)/[P(|F_{L1}|) + P(|F_{L2}|)] \qquad (6.30)$$

while that for $\Delta\phi$ to be negative is

$$P(-\Delta\phi) = P(|F_{L2}|)/[P(|F_{L1}|) + P(|F_{L2}|)], \qquad (6.31)$$

where the appropriate probability distribution, (6.28) or (6.29), is used in (6.30) and (6.31). A Fourier map is then calculated using both possible phases $\phi' + \Delta\phi$ and $\phi' - \Delta\phi$, with Fourier coefficients weighted respectively with $P(+\Delta\phi)$ and $P(-\Delta\phi)$. Such a Fourier map resolves the phase ambiguity. The method was tested on the experimental OAS data of the protein aPP, which is the same test structure as was used in §6.4.5. A result of quality similar to that found from the direct method described previously was obtained.

6.4.7 Use of P_s-function-related methods

Okaya, Saito and Pepinsky (1955) proposed the use of anomalous X-ray dispersion for the solution of crystal structures through interpretation of the P_s function, which is defined as the imaginary part of the Patterson function in the presence of anomalous scattering, i.e.

Fig. 6.5 Comparison of maps from three different phasing methods: (a) averaged OAS, (b) combined OAS and Sim distributions, and (c) combined OAS, Cochran and Sim distributions. The maps are projected along the y axis for the region between fractional coordinates $y=0.11$ and $y=0.21$ and are plotted for $x=0$ to $x=0.5$ and $z=-0.5$ to $z=0.5$. The first contour level of all three maps is about one and a half times the average background. Molecular skeletons are plotted as balls and sticks. Two strong peaks are labelled as Hg and Zn. Note that Hg has a true position of $y=0$, outside the region of this plot while Zn ($y=0.16$) is at the centre of the y region of the plot.

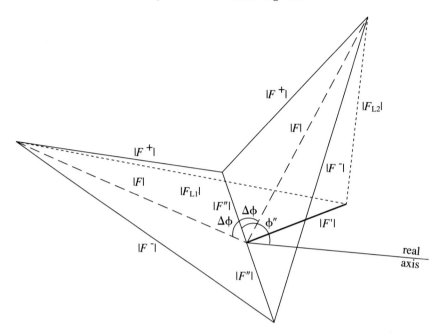

Fig. 6.6 The OAS ambiguity expressed as two possible values for the contribution of the unknown part of the structure. $|F^+|$ and $|F^-|$ are the structure-factor magnitudes of the Friedel pair, $|F|$ is the normal part of the scattering from all atoms and $|F'|$ and $|F''|$ are the contributions of the real and imaginary parts of the anomalous scattering. The two possible contributions of the light (unknown) atoms are $|F_{L1}|$ and $|F_{L2}|$.

$$P_s(\mathbf{u}) = \sum_{\text{for all } h} |F(\mathbf{h})|^2 \sin(2\pi\mathbf{h}\cdot\mathbf{u}) = \sum_{h>0} [|F(\mathbf{h})|^2 - |F(-\mathbf{h})|^2] \sin(2\pi\mathbf{h}\cdot\mathbf{u}). \quad (6.32)$$

The scattering factor of an atom in the case of anomalous scattering can be written as $f = {}_0f' + if'$, where the real part, ${}_0f'$, includes an anomalous component. The structure factor is then written as

$$F(\mathbf{h}) = \sum_{j=1}^{N} ({}_0f_j' + if_j'') \exp(2\pi i \mathbf{h}\cdot\mathbf{r}_j) \quad (6.33)$$

and

$$|F(\mathbf{h})|^2 = F(\mathbf{h})\cdot F^*(\mathbf{h}) = \sum_{j=1}^{N}\sum_{k=1}^{N} ({}_0f_j' + if_j'')({}_0f_k' - if_k'') \exp[2\pi i \mathbf{h}\cdot(\mathbf{r}_j - \mathbf{r}_k)]. \quad (6.34)$$

Similarly

$$|F(-\mathbf{h})|^2 = \sum_{j=1}^{N} \sum_{k=1}^{N} (_0f_j' + if_j'')(_0f_k' - if_k'') \exp[-2\pi i \mathbf{h} \cdot (\mathbf{r}_j - \mathbf{r}_k)]. \quad (6.35)$$

From (6.34) and (6.35)

$$|F(\mathbf{h})|^2 - |F(-\mathbf{h})|^2 = i2 \sum_{j=1}^{N} \sum_{k=1}^{N} (_0f_j' + if_j'')(_0f_k' - if_k'') \sin[2\pi \mathbf{h} \cdot (\mathbf{r}_j - \mathbf{r}_k)]. \quad (6.36)$$

Combining terms involving (j,k) and (k,j) gives

$$(_0f_j' + if_j'')(_0f_k' - if_k'') \sin[2\pi \mathbf{h} \cdot (\mathbf{r}_j - \mathbf{r}_k) + (_0f_k' + if_k'')(_0f_j' - if_j'')$$
$$\times \sin[2\pi \mathbf{h} \cdot (\mathbf{r}_k - \mathbf{r}_j)$$
$$= i2(f_j'' \, _0f_k' - f_k'' \, _0f_j') \sin[2\pi \mathbf{h} \cdot (\mathbf{r}_j - \mathbf{r}_k)]. \quad (6.37)$$

Combining (6.36) and (6.37), we obtain

$$|F(\mathbf{h})|^2 - |F(-\mathbf{h})|^2 = 2 \sum_{j=1}^{N} \sum_{k=1}^{N} (f_k'' \, _0f_j' - f_j'' \, _0f_k') \sin[2\pi \mathbf{h} \cdot (\mathbf{r}_j - \mathbf{r}_k)]. \quad (6.38)$$

Thus the P_s function has peaks of weight proportional to $(f_j'' \, _0f_k' - f_k'' \, _0f_j')$ at position $\mathbf{r}_k - \mathbf{r}_j$ (see, for example, Pepinsky, Robertson and Speakman (1961), pp. 273–7). If all anomalous scatterers are of the same kind then in the P_s function there will be no Patterson peaks relating two anomalous scatterers or two non-anomalous scatterers, since in both cases the value of $(f_j'' \, _0f_k' - f_k'' \, _0f_j')$ equals zero. Because the vector $\mathbf{r}_k - \mathbf{r}_j$ is from atom j to atom k, vectors from anomalous scatterers to non-anomalous scatterers produce positive peaks while vectors in the reverse direction yield negative peaks. The information in the P_s function, considering the positive region alone, consists of a degraded superposition of m images of the structure, each image having a different anomalous scatterer at the origin. The degradation is due to the cancellation of positive and negative peaks.

Pepinsky (1956) and Pepinsky and Okaya (1957) proposed that a superposition method could be used to find a single image of the structure when the positions of the anomalous scatterers were known, and this was demonstrated to be so for small structures. A number of procedures have been proposed for applying the P_s function to the structure solution of proteins. Hao and Woolfson (1989) showed that a superposition of the P_s function calculated from the OAS data can reveal the structure of the protein aPP (the test sample described in §6.4.5). The P_s map was superimposed on the four Hg sites via a sum function (equation (2.12);

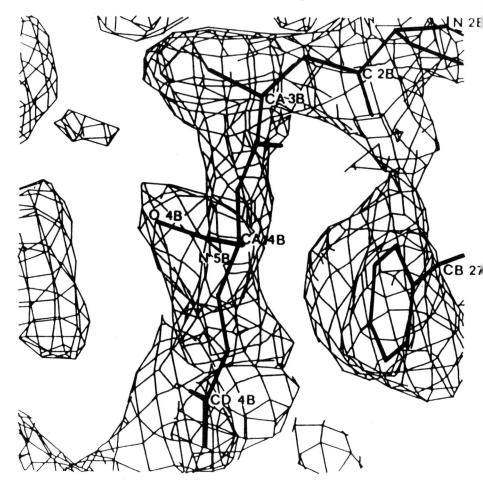

Fig. 6.7 Part of the electron-density map of aPP, calculated with phases derived from the resultant map found by superposition of the P_s map. Appropriate electron density has been inserted at Hg sites.

Buerger, 1959) and negative densities were eliminated from the resultant map. This map, with appropriate density inserted at Hg sites, closely resembles a map calculated with true phases; the two maps have a correlation coefficient of 0.67. For 2108 reflections the unweighted mean phase error is 39.9° but with $|F_o F_c|$ weighting, where F_c is the Fourier coefficient of the final map, this reduces to 29.5°. Part of the electron density map obtained is shown in fig. 6.7, which is of similar quality to the map in fig. 6.4.

Fan, Hao and Woolfson (1990b) proposed an analytical approach, in which the phase, $\phi(\mathbf{h})$, of a structure factor can be determined from the following equation:

$$\tan[\phi(\mathbf{h}) - \varepsilon(\mathbf{h})] = \frac{|F(\mathbf{h})|^2 - |F(-\mathbf{h})|^2}{\chi(\mathbf{h}) + 4g[a(\mathbf{h})^2 + b(\mathbf{h})^2]}, \qquad (6.39)$$

where $\tan[\varepsilon(\mathbf{h})] = b(\mathbf{h})/a(\mathbf{h})$;

$$b(\mathbf{h}) = \sum_{j=1}^{N_{ano}} f_j'' \sin(2\pi\mathbf{h}\cdot\mathbf{r}_j); \qquad a(\mathbf{h}) = \sum_{j=1}^{N_{ano}} f_j'' \cos(2\pi\mathbf{h}\cdot\mathbf{r}_j);$$

N_{ano} is the number of anomalous scatterers in the unit cell, all of the same kind; $\chi(\mathbf{h})$ is calculated from the Fourier transform of the function $|P_s(\mathbf{u})|$; and $g = f'/f''$.

The method has been tested on two protein structures. The first was the Hg-derivative of aPP (§6.4.5). The second was a Pt-derivative of ribonuclease Sa (RNase) (Dodson, Sevcik, Dodson and Zelinka, 1987), crystallising in space group $P2_12_12_1$, with $a = 64.85$ Å, $b = 78.56$ Å and $c = 39.51$ Å. There are two molecules, each with 96 amino-acid residues, in the asymmetric unit. The resultant phases for aPP give an electron density map, which can easily be interpreted in terms of a model (see fig. 6.8). For RNase the map, fig. 6.9, is less clear but has strong similarities with the true map and could probably be interpreted.

Ralph and Woolfson (1991) proposed an alternative procedure for making use of the P_s function. This estimates the magnitudes of the structure factors for the unknown part of the structure and compares the estimate with the values of $|F_{L1}|$ and $|F_{L2}|$ (see fig. 6.6). The estimation is based on the following property of the P_s function. For a structure with only one type of anomalous scatterer, it is possible to calculate an antisymmetric map, the P_s map, which shows positive peaks for vectors from anomalous scatterers to non-anomalous scatterers and negative peaks for vectors in the reverse direction. Now consider the $|P_s|$ map, then this will contain positive peaks for vectors in both directions. In other words the $|P_s|$ map contains positive peaks for vectors between anomalous scatterers and non-anomalous scatterers. By Fourier transformation we can find the Fourier coefficients of this map, $\chi(\mathbf{h})$. On the other hand, a Patterson map with Fourier coefficients $|F|^2$ (see fig. 6.6) would show interatomic vectors between all atoms, while the Patterson map calculated with coefficients $|F'|^2$ (see fig. 6.6) would give vectors between anomalous scatterers only. It is clear that

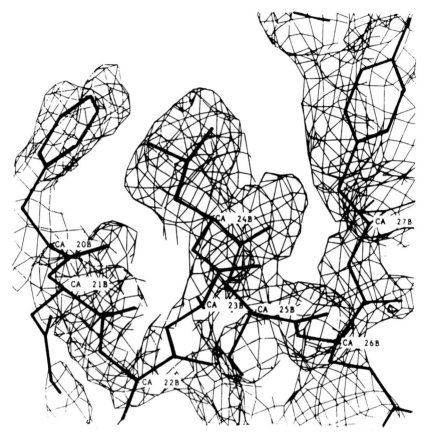

Fig. 6.8 Part of the electron-density map for aPP calculated with phases derived from the analytical method. The true model is superimposed.

vectors between non-anomalous scatterers only
   = vectors between all atoms
      − vectors between anomalous scatterers only
      − vectors between anomalous and non-anomalous scatterers,

which leads to

$$|F_L(\mathbf{h})|^2 = |F(\mathbf{h})|^2 - |F'|^2 - k\chi(\mathbf{h}). \tag{6.40}$$

The constant k is a scale factor and may be calculated from

$$k^2 = \overline{\sum_{j=1}^{m} \sum_{k=1}^{n} f_k^2 (f_j + f_j')^2}^h \Big/ \overline{\chi(\mathbf{h})^2}^h, \tag{6.41}$$

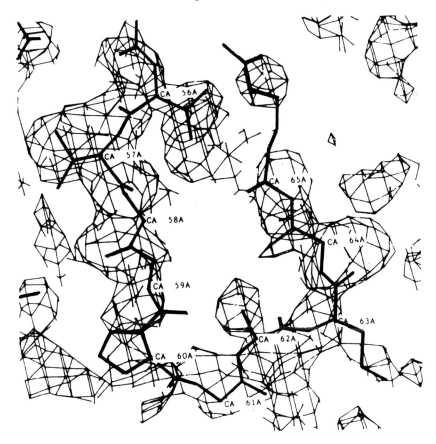

Fig. 6.9 Part of the electron-density map for RNase calculated with phases derived from the analytical method. The true model is superimposed.

where m is the number of anomalous scatterers and n the number of non-anomalous scatterers. The factor k is a correction factor to allow for the elimination of peaks by positive–negative cancellation in the P_s map. The value of $|F_L|$ estimated from (6.40) can be compared to $|F_{L1}|$ and $|F_{L2}|$ and weights for calculating the Fourier map (see §6.4.6) can be deduced as follows, where it is assumed that $|F_{L1}| > |F_{L2}|$.

$$W^+ = (|F_L| - |F_{L2}|)/(2|F_L| - |F_{L1}| - |F_{L2}|) \text{ for } |F_{L2}| > |F_L| \text{ or } |F_L| > |F_{L1}|$$

$$W^+ = (|F_L| - |F_{L2}|)/(|F_{L1}| - |F_{L2}|) \text{ for } |F_{L1}| > |F_L| > |F_{L2}| \qquad (6.42)$$

with $W^- = 1 - W$. A test of the method on aPP gave results similar to that from other P_s-function related methods.

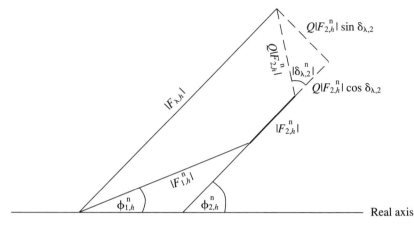

Fig. 6.10 The decomposition of the structure factor with the notation used by Karle (1980).

Methods described in this and previous sections use the same basic data and give mean phase errors of similar magnitude. However, there are significant differences in the distribution of the error and this leads to the possibility of combining two different techniques to obtain something better than either of them individually. Fan, Hao and Woolfson (1990a) showed that the results both from direct methods (§6.4.5) and from the P_s function (§6.4.7) can be improved by simply combining the resultant maps via a minimum function (equation (2.15); Buerger, 1959).

6.5 Developments in multi-wavelength anomalous scattering methods

6.5.1 The MAD technique

The first important development of a method to use multi-wavelength anomalous scattering was the so-called MAD (multi-wavelength anomalous dispersion) technique by Karle (1980). The basis of this method can be followed by reference to fig. 6.10 where the notation used by Karle is used. It is assumed that there is only one type of anomalous scatterer in the structure. The anomalous structure factor is $F_{\lambda,h}$, which has several components. The first of these is $F_{1,h}^n$, which is the contribution of the non-anomalous scatterers; the phase associated with this contribution is $\phi_{1,h}^n$. The second, $F_{2,h}^n$, is the normal scattering from the anomalous scatterers with its associated phase $\phi_{2,h}^n$. The quantity Q is defined by

$$Q = \left[(f_{\lambda,2}')^2 + (f_{\lambda,2}'')^2 \right]^{\frac{1}{2}} / f_{2,h}^n \qquad (6.43)$$

where $f_{2,h}^n$ is the normal part, and $f_{\lambda,2}'$ and $f_{\lambda,2}''$ are the real and imaginary parts, of the anomalous scattering factor. With

$$\tan(\delta_{\lambda 2}) = f_{\lambda,2}''/f_{\lambda,2}', \tag{6.44}$$

then the magnitudes of the real and imaginary parts of the contribution of the anomalous scatterers are $Q|F_{2,h}^n|\cos\delta_{\lambda,2}$ and $Q|F_{2,h}^n|\sin\delta_{\lambda,2}$, as shown in fig. 6.10. By some manipulation of formulae from various triangles that can be formed in fig. 6.10 it can be shown that

$$
\begin{aligned}
|F_{\lambda,h}|^2 = |F_{1,h}^n|^2 &+ [1 + Q(Q + 2\cos\delta_{\lambda 2})]|F_{2,h}^n|^2 \\
&+ 2(1 + Q\cos\delta_{\lambda 2})|F_{1,h}^n||F_{2,h}^n|\cos(\phi_{1,h}^n - \phi_{2,h}^n) \\
&+ 2Q\sin\delta_{\lambda 2}|F_{1,h}^n||F_{2,h}^n|\sin(\phi_{1,h}^n - \phi_{2,h}^n).
\end{aligned}
\tag{6.45}
$$

A similar equation can be produced with $|F_{\lambda,\bar{h}}|$ on the left-hand side where the last term on the right-hand side of (6.45) is preceded by a negative sign. Karle took the variables in this equation as $|F_{1,h}^n|^2$, $|F_{2,h}^n|^2$, $|F_{1,h}^n||F_{2,h}^n|\cos(\phi_{1,h}^n - \phi_{2,h}^n)$ and $|F_{1,h}^n||F_{2,h}^n|\sin(\phi_{1,h}^n - \phi_{2,h}^n)$. There are two equations for each wavelength and in addition there can be added the constraint that

$$
[|F_{1,h}^n||F_{2,h}^n|\cos(\phi_{1,h}^n - \phi_{2,h}^n)]^2 + [|F_{1,h}^n||F_{2,h}^n|\sin(\phi_{1,h}^n - \phi_{2,h}^n)]^2 \\
= |F_{1,h}^n|^2|F_{2,h}^n|^2.
\tag{6.46}
$$

If data are taken at two or more wavelengths then there are more equations than unknowns for each **h** and a least squares solution can be found. The information that is required to solve the structure is the value of $|F_{1,h}^n|$ and its associated phase $\phi_{1,h}^n$. These can only be determined if the value of $\phi_{2,h}^n$ is previously known, which requires a knowledge of the positions of the anomalous scatterers. This topic was previously considered in §4.2.4.

The effectiveness of this approach, or something very similar to it, was well illustrated by the solution of the structure of the protein selenobiotinyl streptavidin (Hendrickson, Pähler, Smith, Satow, Merritt and Phizacker-ley, 1989). The crystals are in space group I222 with unit cell dimensions $a = 95.277$ Å, $b = 105.41$ Å and $c = 47.56$ Å. No structural precedent was available in this study. Three different wavelengths (0.9000, 0.9795 and 0.9809 Å respectively) were used with the selenium atoms serving as the anomalous scatterers. A least-squares procedure (Hendrickson, 1985; Hendrickson, Smith, Phizackerley and Merritt, 1988) was used to find the phases needed for the Fourier synthesis which led to an unambiguous chain tracing at 3.3 Å resolution. A section of the map obtained is shown in fig. 6.11.

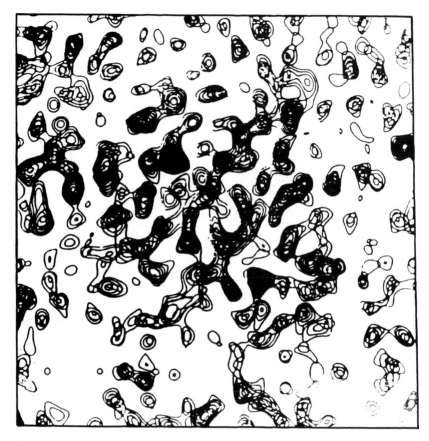

Fig. 6.11 Section of the electron-density map for selenobiotinyl streptavidin calculated with phases derived from the least-squares method for multi-wavelength anomalous scattering.

6.5.2 *Refining observed magnitudes with multi-wavelength data*

If there is only one kind of anomalous scatterer in the structure then it can be shown that Bijvoet intensity differences for different wavelengths must satisfy certain equality relationships in order to be consistent. The analysis that follows will be related to fig. 6.12, where F is the total non-anomalous contribution, including that from the anomalous scatterers, and g is the total anomalous contribution. Then

$$|F^+|^2 = |F|^2 + g^2 + 2|F|g\cos(\theta + \delta), \qquad (6.47a)$$

$$|F^-|^2 = |F|^2 + g^2 + 2|F|g\cos(\theta - \delta). \qquad (6.47b)$$

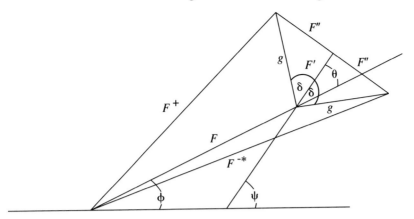

Fig. 6.12 The decomposition of F^+ and F^- with the notation used by Woolfson, Yao and Fan (1993).

Subtracting

$$|F^-|^2 - |F^+|^2 = 4|F|g\sin\theta\sin\delta. \qquad (6.48)$$

Since the values of $|F|$ and θ are wavelength independent and

$$g\sin\delta = |F''| \propto f''$$

then

$$\frac{|F^-|^2 - |F^+|^2}{f''} = C \qquad (6.49)$$

is independent of wavelength. This result was first given by Karle (1984). The variations in the determined values of C for different wavelengths depends on the quality of the data.

Woolfson, Yao and Fan (1993) showed that (6.49) could be used to modify anomalous scattering data to make them self-consistent and that such modified data give improved results when used in phasing procedures. From an anomalous scattering experiment there are usually found the magnitudes of the Friedel pairs $|F^+|$ and $|F^-|$ together with standard deviations σ^+ and σ^-. We shall consider quantities x^+ and x^-, thought of as corrections to the observed structure moduli, such that

$$C_i = [(|F_i^+| + x_i^+)^2 - (|F_i^-| + x_i^-)^2]/f'' \qquad (6.50)$$

has the same value for each wavelength, λ_i, under the condition

$$G = \sum_{i=1}^{n} \left[\left(\frac{x_i^+}{\sigma_i^+} \right)^2 + \left(\frac{x_i^-}{\sigma_i^-} \right)^2 \right] \tag{6.51}$$

is a minimum, where there are observations at n different wavelengths. This gives the maximum joint probability of the set of values of x under the condition that the values of C are all the same. Taking $n = 3$ then from the consistency condition

$$x_2^- = \left\{ \left(|F_2^+| + x_2^+ \right)^2 - \frac{f_2''}{f_1''} \left[\left(|F_1^+| + x_1^+ \right)^2 - \left(|F_1^-| + x_1^- \right)^2 \right] \right\}^{1/2} - |F_2^-|,$$

$$x_3^- = \left\{ \left(|F_3^+| + x_3^+ \right)^2 - \frac{f_3''}{f_1''} \left[\left(|F_1^+| + x_1^+ \right)^2 - \left(|F_1^-| + x_1^- \right)^2 \right] \right\}^{1/2} - |F_3^-|. \tag{6.52}$$

Substituting these values in (6.51) converts the problem to that of minimising $G(x_1^+, x_1^-, x_2^+, x_3^+)$ without constraint. Writing $X_1 = x_1^+$, $X_2 = x_1^-$, $X_3 = x_2^+$, and $X_4 = x_3^+$, the minimum for G is found by iterative application of

$$\Delta X = A^{-1} \times b$$

where the ith element of ΔX, ΔX_i, is the change in X_i, the ith element of b is $-\partial G/\partial X_i$ and $A_{ij} = \partial^2 G/(\partial X_i \partial X_j)$.

Table 6.1 shows the effect of applying this process to two reflections for selenobiotinyl streptavidin at the three wavelengths given in §6.5.1. For (8 6 2), the observed signs of the anomalous differences were all the same but this was not so for (8 10 2). A change of magnitudes by, at most, one standard deviation gives modified magnitudes all giving the same value of C. This process is called REVISE; for some phasing procedures better results are obtained using the REVISE-modified magnitudes rather than the original ones.

6.5.3 The AGREE method

Woolfson, Yao and Fan (1993) devised two effective procedures for obtaining phases from anomalous scattering data. The analysis for the first of these, called AGREE, is now given.

From equations (6.47a, b) by addition and subtraction

$$|F^-|^2 - |F^+|^2 = 4|F|g \sin\theta \sin\delta, \tag{6.53}$$

$$|F^-|^2 + |F^+|^2 = 2|F|^2 + 2g^2 + 4|F|g \cos\theta \cos\delta, \tag{6.54}$$

Table 6.1. *The effect of applying the REVISE process to observed structure factors at three wavelengths for the structure selenobiotinyl streptavidin*

	$\lambda = 0.9000$ Å		$\lambda = 0.9795$ Å		$\lambda = 0.9809$ Å	
	h	h̄	h	h̄	h	h̄
Reflection (8 6 2)						
Experimental $|F|$	917.1	958.3	949.5	989.9	942.6	976.3
σ	8.25	8.37	8.78	8.58	6.66	6.60
Value of C	$-23\,544$		$-38\,063$		$-17\,681$	
Modified $|F|$	918.7	956.6	958.3	981.3	938.9	980.2
Revised C	$-21\,653$		$-21\,653$		$-21\,653$	
Reflection (8 10 2)						
Experimental $|F|$	229.0	178.6	237.7	243.4	240.2	204.4
σ	20.06	24.48	19.29	17.71	14.84	18.03
Value of C	-6254		758		-7742	
Modified $|F|$	221.8	187.7	251.0	233.0	239.8	204.9
Revised C	-4242		-4242		-4242	

Eliminating θ gives

$$Pg^4 + Qg^2 + R = 0, \qquad (6.55)$$

where

$$P = 4 \sin^2 \delta,$$
$$Q = -4[|F^-|^2 + |F^+|^2 - 2|F|^2] \sin^2 \delta - 4|F|^2 \sin^2 (2\delta),$$
$$R = |F^+|^4 + |F^-|^{+4} - 2|F^+|^2|F^-|^2 \cos(2\delta) - 4|F|^2 \sin^2 \delta (|F^-|^2 + |F^+|^2 - |F|^2)$$

and δ is defined in (6.44).

If $|F|$ were known then the ambiguity associated with anomalous scattering would reside in the magnitude of g since from (6.55)

$$g^2 = [-Q \pm (Q^2 - 4PR)^{\frac{1}{2}}]/2P. \qquad (6.56)$$

Given

$$g^2 = |F'|^2 + |F''|^2$$

$$F'(\mathbf{h}) = f' \sum_{j=1}^{M} \exp(2\pi i \mathbf{h} \cdot \mathbf{r}_j),$$

$$F''(\mathbf{h}) = f'' \sum_{j=1}^{M} \exp(2\pi i \mathbf{h} \cdot \mathbf{r}_j)$$

for M anomalous scatterers at positions $\mathbf{r}_j$, $j = 1$ to M, it follows that

$$m^2 = \frac{g^2}{(f')^2 + (f'')^2} = \left| \sum_{j=1}^{M} \exp(2\pi i \mathbf{h} \cdot \mathbf{r}_j) \right|^2$$

is independent of wavelength.

For a wavelength λ_i the maximum value of $|g_i|$ is

$$|g_i|_{\max} = M[(f_i')^2 + (f_i'')^2]^{\frac{1}{2}}. \tag{6.57}$$

These maximum values, together with geometrical constraints, lead to the inequalities, applicable for any wavelength, λ_i

$$|F| < |F_i^+| + |g_i|_{\max}$$
$$|F| < |F_i^-| + |g_i|_{\max}$$
$$|F| > |F_i^+| - |g_i|_{\max}$$
$$|F| > |F_i^-| - |g_i|_{\max} \tag{6.58}$$

The AGREE method explores the possible range of $|F|$ values, at a number of equi-interval points over the range, and compares the values of m_i^2 found for each wavelength. The consistency of these is found from either

$$T = \left(\sum_{i=1}^{M-1} \sum_{j=i+1}^{M} (m_i^2 - m_j^2)^2 \right)^{1/2} \tag{6.59a}$$

or a scaled version

$$T' = \frac{T}{\Sigma m_i^2} \tag{6.59b}$$

and the minimum value of T, or T', indicates the most probable values of m and hence g, and also $|F|$. The value of θ is then found from

$$\tan \theta = \frac{(|F^-|^2 - |F^+|^2)\cos \delta}{(|F^+|^2 + |F^-|^2 - 2|F|^2 - 2g^2)\sin \delta}, \tag{6.60}$$

where the ambiguity in the value of θ is resolved by the signs of the numerator and divisor of the expression.

It is evident that the analysis leading to the values of g and $|F|$ has not required any information about the positions of the anomalous scatterers, but only their total number and type. Actually the analysis has given values of g, which, if they are good enough, can give by the use of a Patterson function with Fourier coefficients g^2 (or m^2) or with MULTAN with structure amplitudes g (or m), the positions of the anomalous scatterers. From these positions F' can be found in both magnitude and phase (ψ) and hence the phase of F from

$$\phi = \psi - \theta$$

(see fig. 6.12).

In table 6.2 are shown the results of the AGREE process for two different reflections for selenobiotinyl streptavidin. For reflection (8 2 2) there is a well-defined minimum of T' and the three values of m^2 agree quite well. For the (9 7 2) reflection the minimum of T' is more poorly defined and there is also less agreement of the values of m^2 at the minimum. Most situations fall between these two extremes.

It is found that smaller values of T' correspond to more reliable estimates of m^2: table 6.3 shows the values of the residual

$$R = \frac{\sum |m_{est}^2 - m_{true}^2|}{\sum m_{true}^2} \tag{6.61}$$

for batches of reflections in ranges of values of T', where m_{est}^2 and m_{true}^2 are the estimated and true values of m^2 and the sums are over all reflections in the batch. Smaller values of T' give better estimates of m although even the best agreement does not look impressive. Nevertheless it is found that the estimated values of m^2 are actually good enough to give useful information.

The combined use of REVISE and AGREE can be very effective. Table 6.4 shows the top 20 peaks in a Patterson map for selenobiotinyl streptavidin, with coefficients $\overline{m^2}$ for the 2000 most reliably indicated values as indicated by T' after the data have been modified by REVISE. The ten interatomic peaks between the selenium atoms are present and, given that the correct peaks are a related set, they can readily be recognised.

Having found the positions of the anomalous scatterers from values of $|F|$ and g determined by AGREE, it is possible to find estimated phases for all 4217 reflections of selenobiotinyl streptavidin, including the centric ones. The estimated values of ϕ, using data unmodified by REVISE give a mean phase error of 62.0° with a corresponding map correlation coefficient of 0.456; the map contains regions that can be readily interpreted. From the data modified by REVISE there is a substantially better result although only 4018 reflections are available since some reflections failed to refine properly with our REVISE procedure. The mean phase error is 54.3° and the map correlation coefficient is 0.549. This shows the benefit of subjecting

Table 6.2. *Values of* $|F|$, m_i^2, $\overline{m^2}$ *and* T' *for two reflections for values of* $|F|$ *in the vicinity of the minimum of* T' *together with the values of* $|F^+|$ *and* $|F^-|$ *for the three wavelengths.*

	Reflection (8 2 2)			$|F^+|$	$|F^-|$	
			λ_1	985.8	993.2	
			λ_2	1042.2	1049.5	
			λ_3	1021.7	1031.3	

$|F|$	m_1^2	m_2^2	m_3^2	$\overline{m^2}$	T'
973.99	87.69	78.21	72.06	79.31	0.0813
974.37	83.71	77.42	71.03	77.39	0.0669
974.74	79.82	76.63	70.04	75.50	0.0539
975.12	76.02	75.84	69.06	73.64	0.0440
975.49	72.30	75.06	68.09	71.82	0.0399
975.86	68.68	74.28	67.13	70.03	0.0439
976.24	65.15	73.51	66.17	68.28	0.0546
976.61	61.71	72.74	65.23	66.56	0.0692
976.99	58.36	71.98	64.28	64.87	0.0859
977.36	55.10	71.22	63.35	63.22	0.1041
977.74	51.93	70.46	62.42	61.60	0.1231

	Reflection (9 7 2)			$|F^+|$	$|F^-|$	
			λ_1	564.8	587.0	
			λ_2	556.7	604.7	
			λ_3	555.9	579.0	

$|F|$	m_1^2	m_2^2	m_3^2	$\overline{m^2}$	T'
591.63	117.31	136.15	24.99	92.82	0.5233
591.84	120.37	136.17	25.26	93.93	0.5215
592.06	123.48	136.19	25.53	95.07	0.5201
592.27	126.64	136.21	25.80	96.22	0.5190
592.48	129.85	136.24	26.08	97.39	0.5184
592.69	133.11	136.27	26.36	98.58	0.5182
592.90	136.43	136.29	26.64	99.79	0.5183
593.11	139.79	136.32	26.93	101.01	0.5188
593.32	143.21	136.35	27.21	102.26	0.5197
593.53	146.67	136.38	27.50	103.52	0.5208
593.74	150.19	136.42	27.79	104.80	0.5224

Table 6.3. *The residual (6.61) for the* N *most reliably indicated values of* m^2 *according to the values of* T', *found from REVISE and AGREE. The final row, for* N = 4018, *gives the overall residual for all available reflections*

N	Residual
500	0.744
1000	0.783
1500	0.790
2000	0.811
2500	0.837
3000	0.862
3500	0.880
4000	0.928
4018	0.946

Table 6.4. *The top 20 peaks, excluding the origin peak, from a map with the 2000 most reliably indicated values of* m^2, *as indicated by* T' *following the use of REVISE and AGREE, used as coefficients. The ten peaks between selenium atoms are numbered*

Peak	Height	x	y	z	Peak no.
1	530	0.0000	0.0997	0.5000	1
2	480	0.4177	0.0050	0.5000	2
3	402	0.0000	0.0526	0.5000	
4	390	0.1330	0.0622	0.3396	3
5	363	0.1312	0.1671	0.1472	4
6	343	0.5000	0.0023	0.4333	
7	342	0.3140	0.0030	0.1693	5
8	335	0.0000	0.0205	0.0798	
9	335	0.4239	0.1090	0.0000	6
10	329	0.5000	0.1229	0.5000	
11	322	0.3321	0.2220	0.0000	7
12	311	0.0216	0.0395	0.0000	
13	305	0.0000	0.2287	0.1900	8
14	293	0.2098	0.0004	0.4991	
15	280	0.0425	0.0006	0.0485	
16	273	0.4492	0.0649	0.1526	9
17	262	0.4706	0.1869	0.5000	
18	254	0.5000	0.0961	0.3304	
19	253	0.4556	0.1630	0.3389	10
20	252	0.0000	0.1550	0.0404	

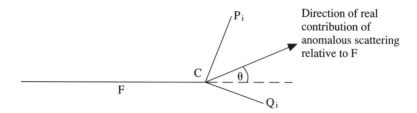

Fig. 6.13 F is the contribution of the non-anomalous scattering for all atoms and CP_i and CQ_i, with equal magnitudes, are the contributions of the anomalous scattering for wavelength λ_i to F^+ and F^-.

the values of $|F^+|$ and $|F^-|$ to REVISE before using them in any phase-estimating process.

6.5.4 The ROTATE method

The second of the methods given by Woolfson, Yao and Fan (1993) is called ROTATE and this is now described. In fig. 6.13 are shown anomalous structure amplitudes turned so that the non-anomalous contribution, F, is drawn horizontally and the anomalous contributions for λ_i are CP_i and CQ_i for F^+ and F^- respectively. The angle θ is, as in fig. 6.12, that between F and the real contribution of the anomalous scattering, and is wavelength-independent. We have

$$(CP_i)^2 = (CQ_i)^2 = m^2[(f_i')^2 + (f_i'')^2], \qquad (6.62)$$

where the value of m (equation (6.56)) comes from the assumed known positions of the anomalous scatterers. If values of $|F|$ and θ were known then it would be simple to calculate the values of OP_i and OQ_i, which, in a perfect situation, would equal $|F_i^+|$ and $|F_i^-|$. If some assumed values $|F|$ and θ are considered then one may write

$$\Delta|F_i^+| = OP_i - |F_i^+|,$$
$$\Delta|F_i^-| = OQ_i - |F_i^-|,$$
$$S(|F|, \theta) = \sum_i \left[\left(\frac{\Delta|F_i^+|}{\sigma_i^+} \right)^2 + \left(\frac{\Delta|F_i^-|}{\sigma_i^-} \right)^2 \right], \qquad (6.63)$$

where the summation is over all available wavelengths λ_i.

Starting with values of $|F|$ and θ that are not too far from the correct values, improved estimates of their values can be obtained by minimisation of S. This was done by a steepest descents approach with successive shifts in $|F|$ and θ given by

$$\delta\theta = \frac{-\alpha S \dfrac{\partial S}{\partial\theta}}{\left[\left(\dfrac{\partial S}{\partial\theta}\right)^2 + \left(\dfrac{\partial S}{\partial|F|}\right)^2\right]^{1/2}}, \tag{6.64a}$$

$$\delta|F| = \frac{-\alpha S \dfrac{\partial S}{\partial|F|}}{\left[\left(\dfrac{\partial S}{\partial\theta}\right)^2 + \left(\dfrac{\partial S}{\partial|F|}\right)^2\right]^{1/2}}, \tag{6.64b}$$

where α is a damping constant, which is made small (about 0.01) to avoid the refinement repeatedly jumping over the minimum point.

In a computer programme ROTATE, six different values of θ, the sextant values 30°, 90°, 150°, 210°, 270° and 330° were taken as refinement starting points together with a value of $|F|$ equal to the average of all the available values of $|F^+|$ and $|F^-|$. Some results for the selenobiotinyl streptavidin data are shown in table 6.5. Most situations are as seen for (8 2 2), (8 4 2) and (9 5 2), where all starting points refine to the same final values. Sometimes, as for (9 7 2) and (8 24 2), two different minima are found but normally they are well distinguished in plausibility by the associated values of S.

Without applying REVISE to the data the values of θ led to a mean phase error of 57.0° with a corresponding map correlation coefficient of 0.512. Pre-processing the data with REVISE reduced the phase error to 54.9° with corresponding map correlation coefficient 0.519. A slightly better result than either of these results was obtained by taking that value of θ from the two approaches, with and without using REVISE, which had the smaller value of S. This gave a mean phase error of 54.4° with an MCC of 0.527. The map corresponding to this last result shows the form of most of the molecule; a section of it with part of the molecule superimposed is shown in fig. 6.14.

6.6 Conclusions concerning anomalous scattering methods

The provision of synchrotron radiation X-ray sources is growing rapidly and the most powerful of these sources have, or will have, beamlines dedicated to X-ray crystallography in general and sometimes to anomalous

Table 6.5. *Results from ROTATE for a selection of selenobiotinyl streptavidin data. The final column gives the number of cycles in the refinement, limited to 500*

h k l	θ_{init}	θ_{final}	$|F|_{init}$	$|F|_{final}$	$\overline{|\Delta|F^{\pm}||}$	Number of cycles
8 2 2	30°	157°	1020.6	987.4	11.07	500
	90°	158°		986.9	10.65	500
	150°	157°		987.3	10.99	500
	210°	159°		986.8	10.52	500
	270°	159°		986.5	10.24	500
	330°	159°		986.9	10.57	500
8 4 2	30°	311°	721.2	736.4	7.85	171
	90°	311°		736.3	7.86	162
	150°	305°		734.6	7.82	251
	210°	307°		735.1	7.82	192
	270°	308°		735.5	7.82	166
	330°	307°		735.3	7.82	182
8 24 2	30°	84°	1562.2	1564.0	31.22	19
	90°	4°		1610.9	20.87	242
	150°	85°		1563.5	31.28	30
	210°	3°		1610.8	20.86	271
	270°	7°		1609.9	20.95	214
	330°	6°		1610.7	20.93	236
9 5 2	30°	166°	220.5	204.9	11.66	163
	90°	166°		204.9	11.65	164
	150°	166°		205.0	11.66	163
	210°	165°		204.9	11.65	208
	270°	163°		205.1	11.64	375
	330°	167°		204.9	11.66	204
9 7 2	30°	88°	574.7	573.9	12.74	156
	90°	87°		564.7	12.75	5
	150°	89°		573.8	12.74	41
	210°	99°		566.8	12.84	116
	270°	95°		569.0	12.76	45
	330°	276°		575.3	36.38	22

scattering measurements in particular. Where a structure contains a suitable scatterer it may be as simple to collect anomalous scattering data as a conventional data set and it is evident that the use of anomalous scattering methods should increase at an accelerating rate.

There are obvious attractions in using one-wavelength anomalous scattering because this avoids the necessity of scaling one data set to another, which is a very uncertain procedure. However, for larger struc-

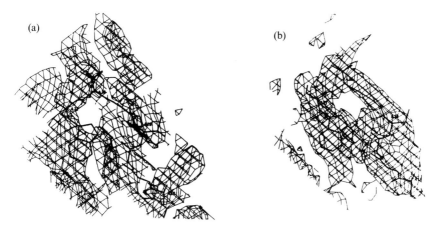

Fig. 6.14(a) A section of a selenobiotinyl streptavidin map calculated with phases from a combination of ROTATE and REVISE + ROTATE with part of the molecule superimposed. (b) The same section of map with true phases calculated from the final refined structure.

tures the one-wavelength methods will work either badly or not at all and then multi-wavelength methods come into their own.

There is available a good theoretical base, which should enable structures of very large size to be solved with multi-wavelength methods but a great deal will depend on the quality of the data and the improvement of this will become a top priority of the experimentalist.

7

Phase extension and refinement

7.1 Introduction

It is easily seen from either the Sayre equation (3.52) or the tangent formula
(3.60) that direct methods are likely to be much more powerful in phase
extension and refinement than in *ab initio* phasing, since nothing can be
known about the left-hand side of either the Sayre equation or the tangent
formula without first putting into the right-hand side at least a small
number of starting phases. One of the reasons why multi-solution procedures are so successful in practice is that they provide the possibility of
having some trial sets with an initial pattern of phases able to converge to
the correct point in the multi-dimensional phase space. On the other hand,
if the phases of a sufficiently large number of reflections can be estimated in
advance then direct methods will work even more efficiently. This gives the
possibility of combining direct methods with other methods to tackle the
phase problem in a number of special cases.

7.2 Fragment development

In the analysis of complex crystal structures, it is often the case that a
fragment instead of the complete structure is first obtained. Hence fragment
development plays an important role in crystal structure determination.

7.2.1 Recycling methods

Fourier synthesis with partial-structure phases has been a very efficient
approach to obtaining the complete structure, especially when this is
associated with weighting functions (Woolfson, 1956; Sim, 1960). A
reciprocal-space alternative is a phase extension and refinement procedure
based on partial structure information. Attempts have been made to use the

Sayre equation to refine partial-structure phases (Hoppe, 1962, 1963; Fan, 1965a; Krabbendem and Kroon, 1971) and these methods have been shown to be reasonably efficient. Alternatively, Karle (1968) proposed a procedure, normally referred to as *Karle recycling*, which makes use of the tangent formula. This method and its successors (Hull and Irwin, 1978; Yao, 1983) have proved, in most cases, to be much more powerful than calculating successive Fourier syntheses.

In Karle recycling a set of phases is calculated from a relatively small structure fragment. Phases are accepted as starting phases if $|F|_{\text{calc}} \geq p|F|_{\text{obs}}$, where p is the fraction of the total scattering power contained in the fragment while $|F|_{\text{obs}}$ is associated with an $|E|_{\text{obs}} \geq 1.5$. As a working rule, often varied, if $p < 0.25$, it is made equal to 0.25 and if $p > 0.6$, it is replaced by 0.6. The starting phases so obtained are input into the tangent formula to calculate a new set of phase estimates. This leads in turn to a new Fourier map, in which a larger fragment of the structure will usually be revealed. The process is iterative; two examples of the successful application of Karle recycling are given here.

1. Example 1. Valinomycin (Karle, 1975) crystallises in space group P1 with two molecules (156 non-hydrogen atoms) in the asymmetric unit. The recycling process was initiated with only nine atoms, corresponding to 6% of the total structure. The acceptance value of p was made equal to 0.10 in this example, a variation on the working rule given above. As the number of atoms found increased so p was correspondingly increased to its maximum value of 0.60. Nine cycles of tangent refinement and a final difference map led to the complete structure.

2. Example 2. The lithium ion complex of antamanide associated with a bromine ion (Karle, 1974) crystallises in space group P2$_1$ with 95 atoms in the asymmetric unit. The recycling was started from the position of the bromine atom. The first result was an E-map containing fragments of both enantiomorphs owing to the centrosymmetric arrangement of the bromine atoms in the unit cell. Fourteen atoms were chosen according to chemical knowledge so as to break the enantiomorphous ambiguity. The complete structure was obtained by iterative tangent refinement and Fourier synthesis.

The Karle recycling procedure, although powerful, is not automatic and the process has to be interrupted at the end of each cycle to interpret the latest E-map and to find more atoms to include in the next cycle. An obvious extension of this method is to assign a weight to each phase and allow the phase to develop and refine as in the programme system

MULTAN (Germain, Main and Woolfson, 1971). For this purpose Hull and Irwin (1978) introduced the probability distribution given by Sim (1960) as a weighting scheme into tangent refinement. This forms a part of the statistically weighted tangent refinement (SWTR) procedure in MUL-TAN-78 (§3.5.3). The main points of the procedure are as follows.

(1) A nucleus of reliably estimated phases is established using the criteria given by Karle (1968).
(2) The probability distribution (2.44) is then used to estimate the reliabi-lity of the phase indication for each E_h given a known fragment, of which the contribution to the normalised structure factor equals e_h. This distribution is of the form

$$P(\varphi_h) = \exp[X_h \cos(\varphi_h - \theta_h)]/2\pi I_0(X_h), \qquad (7.1)$$

where

$$X_h = 2|E_h e_h|/\sigma,$$

$$\sigma = \sum_{p=1}^{N'} z_p^2 / \sum_{j=1}^{N} z_j^2,$$

N' and N are the numbers of atoms in the partial structure and complete structure respectively, θ_h is the calculated phase of the partial structure, and z_j is the atomic number of the jth atom.
(3) Phases from step (1) with $X_h > 4.0$ are used as starting phases for tangent-formula extension and they are given initial weights of $I_1(X_h)/I_0(X_h)$. Two cycles of SWTR are carried out, which involve only the starting phases.
(4) Up to five expansion cycles are performed keeping the starting phases unchanged. The extra reflections at this stage are those from stage (1) for which $X_h > 2.4$ that were not in the starting set. The weights and phases of the extra reflections are changed only at the end of each cycle, but a new phase estimate is only accepted if $\alpha(\mathbf{h})^2 > 2.0$ (see (3.61)). To give extra stability to this stage of the procedure all phases for which the weights become greater than 0.9 are fixed thereafter.
(5) Two cycles of refinement are carried out with all phases being included and allowed to change.
(6) An E-map is calculated excluding all those reflections for which the weight is less than 0.1. All others are given unit weight in the E-map.

SWTR works more reliably and progresses to the final solution more rapidly than does the original Karle recycling process. In practice, even with a small fragment, all or most of the structure is found in a single cycle.

Yao (1983) also improved on Karle recycling by making use of the RANdom starting TANgent refinement (RANTAN) procedure (§3.5.4) (Yao, 1981). A number of the most reliably indicated phases from the Karle criterion are *accepted* with appropriate weights and a large number of other reflections are given random phases with the usual RANTAN weights. This makes fragment development a multi-solution approach, which is quite automatic. Given a good base of starting reflections the number of trials need not be too large, fifty or so is usually sufficient, and the phase set with the best figures of merit will usually show the complete structure, or nearly so.

7.2.2 Tangent refinement applied to difference structure factors

It is known that, in some circumstances, a difference Fourier map can reveal more details of the unknown part of the structure than will an ordinary Fourier map. A direct-method procedure dealing with the 'difference structure factors' (Noordik and Beurskens, 1971; van den Hark, Prick and Beurskens, 1976) is a reciprocal-space analogue and has been proved to be much better than difference Fourier methods, especially when the known part of the structure does not dominate the phases owing either to its relatively small scattering contribution or because of the existence of certain pseudosymmetry. The method, and the corresponding programme system (Beurskens, Bosman, Doesburg, Gould, van den Hark, Prick, Noordik, Beurskens, Parthasarathi, Bruins-Slot, Haltiwanger, Strumpel, Smits, García-Granda, Smykalla, Behm, Schäfer and Admiraal, 1990) is referred to as DIRDIF which stands for 'DIRect methods for DIFference structures'. The main features of the method are given below.

(1) Difference structure factors, F_D, are calculated by subtracting the contribution of the partial structure of known atoms, F_P, from the observed structure factor, F_O, assuming as a first approximation that the phases of the F_O terms are assumed to be the same as those of the F_P terms. If $|F_O|$ is greater than $|F_P|$ then the phase of F_D, ϕ_D, is taken to be the same as that of F_P, ϕ_P, with a weight similar to (2.41) (Woolfson, 1956), or (2.46) (Sim, 1960) or as proposed by Srinivasan (1968). However, if $|F_P| \gg |F_O|$ the ϕ_D is indicated as $\phi_P + \pi$ and the weight is taken as unity.

(2) The values of F_D and ϕ_D are then used as input to a weighted-tangent-formula refinement. As the estimated values of ϕ_D change so the initial approximation $\phi_D = \phi_P$ is invalidated and the estimated amplitudes, $|F_D|$, are changed so that $|F_O| = |F_P + F_D|$.

(3) If the symmetry of the partial structure is higher than that of the whole structure, a symbolic addition technique (Beurskens and Prick, 1981) is introduced to solve the phase ambiguity caused by the pseudosymmetry. As an example to illustrate this feature of DIRDIF we take the situation where the complete structure is non-centrosymmetric but the partial structure, perhaps a single heavy atom, is centrosymmetric. The phases, ϕ_p, and hence also ϕ_D, will be 0 or π and in order to break the centrosymmetric pattern it is necessary to find some reflections with general phases far from these special values. The most likely candidates are those for which $|F_p|$ is small and for which the tangent formula gives a weak phase indication because individual contributions to the tangent formula are inconsistent. A small number of such reflections have their phases indicated by letter symbols and relationships between the symbols are found by symbolic addition (§3.5.2). These relationships are then interpreted on the basis that each symbol is either $\pi/2$ or $-\pi/2$ and the enantiomorph is selected by choosing one of the two possible solutions. Thereafter these phases are incorporated into the tangent formula and refined normally.

DIRDIF is a very useful supplement to conventional direct methods. When a direct method fails to solve a structure, usually due to aberrant phase relationships, it is often possible to recognise a fragment, which can then be used for DIRDIF recycling. In this case the difference structure factor approach offers the likelihood of getting rid of the problem caused by the original aberrant phase relationships, since now all reflections will have different $|E|$ values and the set of Σ_2 relationships will not be the same. In addition, the subtraction of the known part of the structure often also removes the difficulty that may have originated from it.

The efficiency of DIRDIF can be seen from the following examples.

NORA, (Noordik and Beurskens, 1971) $Au(S_2CN(C_4H_9)_2)_2$ $Au(S_2C_2(CN)_2)_2$, belongs to space group $P2_1/c$, with $Z=2$. The structural model is shown in fig. 7.1. The positions of the gold atoms $(0,0,0; \frac{1}{2},0,0; 0,\frac{1}{2},\frac{1}{2}; \frac{1}{2},\frac{1}{2},\frac{1}{2})$ were found by the Patterson method. Since the arrangement of gold atoms gives rise to a sub-periodicity of $t_1 = a/2$ and $t_2 = b/2 + c/2$, reflections with h odd or $k+l$ odd are systematically weak and contain no contribution from the gold atoms. The ordinary heavy-atom method would have resulted in a four-fold positional ambiguity for the remaining atoms. However, DIRDIF, based on the positions of gold atoms, gave the true structure.

MONOS (Noordik, Beurskens, Ottenheijm, Herscheid and Tijhuis, 1978), $C_{15}H_{16}N_2O_2S$, belongs to space group $P2_12_12_1$ with $Z=4$ (fig. 7.2).

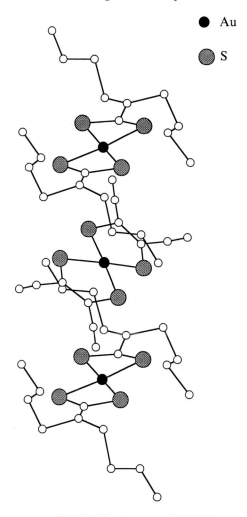

Fig. 7.1 The structure of NORA.

The independent sulphur atom is approximately situated at (0.00, 0.09, 0.14) as found by the Patterson method. This leads to a centrosymmetric arrangement of the sulphur atoms and causes the enantiomorphous ambiguity for the remaining atoms. The structure was solved by DIRDIF in two different ways.

(1) The x-coordinate of the sulphur atom was shifted about 0.15 Å away from 0.00 in the positive direction. Starting with the new sulphur position (0.02, 0.09, 0.14), a comparison was made between the conventional Fourier synthesis and the DIRDIF procedure. Among

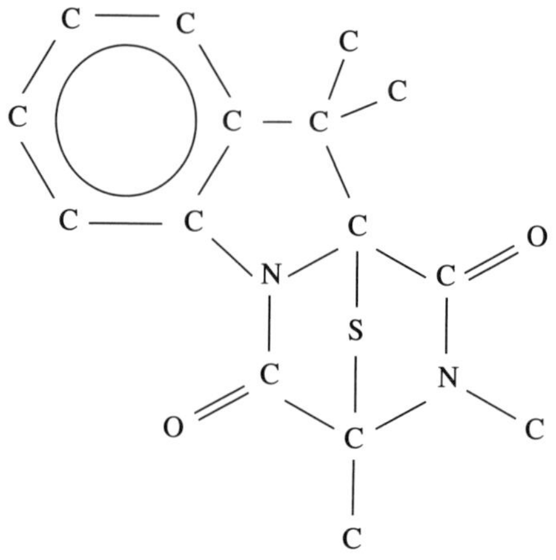

Fig. 7.2 The structure of MONOS.

the twenty-three highest peaks in the resulting peak list, the former technique showed only eleven out of the total twenty non-hydrogen atoms, while DIRDIF showed the complete structure.

(2) The centrosymmetric sulphur atoms were used as input to DIRDIF. The programme recognised the enantiomorphous ambiguity problem. An automatic pre-scan of the inconsistent tangent-formula result was used, as described above, to select ten reflections, which were expected to have phases deviating from 0 or π. These reflections were used to initiate a symbolic phasing procedure, which led to the complete structure.

Other methods dealing with the problems arising from pseudo-symmetry will be discussed in the following two sections.

7.3 Phase extension from real to imaginary parts of structure factors – dealing with pseudo-centrosymmetry

Non-centrosymmetric structures can reveal pseudo-centrosymmetry if they contain some dominating heavy atoms in a centrosymmetric arrangement. In this case either the Patterson method or conventional direct methods will result in a pseudo-centrosymmetric image, in which the true structure is superimposed on its enantiomorph. While this can make difficult the problem of solving the complete structure such an image does contain useful information for phase extension and refinement. By Fourier trans-formation of such an image with respect to an origin chosen at the pseudoinverse centre, both the magnitudes and the signs of the real parts of structure factors can be approximately calculated. Hence the magnitudes of the imaginary parts of structure factors may be obtained from

$$|B(\mathbf{h})| = \left[|F(\mathbf{h})|^2 - A(\mathbf{h})^2\right]^{\frac{1}{2}}, \tag{7.2}$$

where $A(\mathbf{h})$ denotes the real and $B(\mathbf{h})$ the imaginary parts of the structure factor $F(\mathbf{h})$. In the above equation, the sign of $B(\mathbf{h})$ remains unknown, leading to the problem of the enantiomorphous phase ambiguity. In order to solve this problem, we need a special procedure for extending phases (signs) from the real components to the imaginary components of structure factors. This can easily be done by making use of the *component relation* (Fan, 1965b)

$$B(\mathbf{h}) = (2\theta/V)\sum_{\mathbf{h}'} A(\mathbf{h}')B(\mathbf{h}-\mathbf{h}'), \tag{7.3}$$

where θ is an atomic form factor and V the volume of the unit cell. The procedure is illustrated by the following example (Fan and Zheng, 1978).

ZCW, $C_{34}O_{11}NH_{47} \cdot HI$, crystallises in space group $P2_1$ with unit cell parameters $a = 12.58$ Å, $b = 14.38$ Å, $c = 11.00$ Å, $\beta = 114.6°$ and $Z = 2$. There are two iodine atoms centrosymmetrically arranged in the unit cell. RANTAN (Yao, 1981) failed to solve the structure in more than 500 trials. Patterson analysis resulted in a map containing both enantiomorphs together with a few ghost peaks. In addition to the iodine atom, the largest 44 pairs of centrosymmetrically related peaks (among which two pairs turned out to be false peaks) were chosen to estimate the real parts of the structure factors. Estimates for $A(\mathbf{h})$ and $|B(\mathbf{h})|$ were calculated for 331 of the strongest reflections although only 79 $|B(\mathbf{h})|$ terms had estimated values greater than zero. The signs of these $B(\mathbf{h})$ terms were then derived by a symbolic addition procedure based on the component relation (7.3). A Fourier map calculated with the 331 $A(\mathbf{h})$ and 79 $B(\mathbf{h})$ terms revealed unambiguously 33 of the total of 46 light atoms (fig. 7.3a); the final Fourier map is shown in fig. 7.3b. The above method has been modified to a multi-solution approach and incorporated in the programme SAPI (Fan, Yao, Zheng, Gu and Qian, 1991).

7.4 Phase extension from one reflection subset to the others – dealing with pseudo-translational symmetry

An important kind of crystal structure having pseudo-translational symmetry is the so-called modulated structure, in which the atoms suffer from

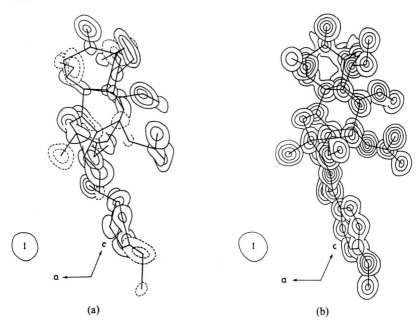

Fig. 7.3 Resolving the enantiomorphous ambiguity for ZCW by making use of the component relation. (a) Fourier map calculated with 331 $A(\mathbf{h})$ terms and 79 $B(\mathbf{h})$ terms. Atoms unambiguously located are denoted by solid contour lines. (b) the final Fourier map.

certain occupational and/or positional fluctuation. Modulated phases have been found in many important inorganic and organic solids. Since, in many cases, the transition to the modulated structure corresponds to a change of physical properties it is important to know the structure of modulated phases in order to understand the mechanism of the transition and the basis for the modification of properties of the modulated state.

A modulated structure can be regarded as the result of applying a periodic modulation to a regular structure. Fig. 7.4 shows two simplified examples. The modulation wave in fig. 7.4 represents the fluctuation of atomic occupancy. When it is applied to the background regular structure, the 'heights' of the atoms are modified. A commensurate modulated structure (superstructure) will result (fig. 7.4a) if the period $\mathbf{T}$ of the modulation function is commensurate with the period $\mathbf{t}$ of the structure, i.e. $\mathbf{T}/\mathbf{t} = n$, where n is an integer. The resulting superstructure now has a true period $\mathbf{T}$ and a pseudo-period $\mathbf{t}$, respectively corresponding to a true unit cell and a pseudo-unit cell. On the other hand, if $\mathbf{T}$ is incommensurate with $\mathbf{t}$ (fig. 7.4b), i.e. $\mathbf{T}/\mathbf{t} = r$, where r is not an integer, we obtain an incommensur-

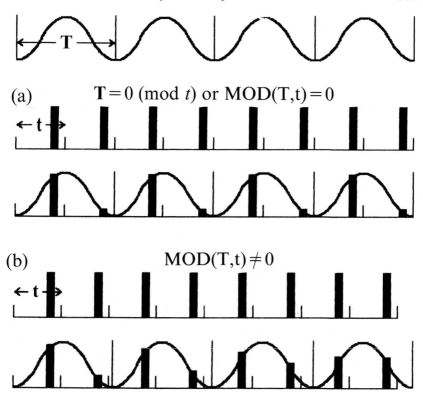

(a) $T = 0 \pmod{t}$ or $MOD(T,t) = 0$

(b) $MOD(T,t) \neq 0$

Fig. 7.4 Occupational modulation of a one-dimensional structure: (a) commensurate modulation, and (b) incommensurate modulation. Top: modulation function with a period equal to **T**; middle: one-dimensional structure with atoms shown as thick vertical lines and with a period equal to **t**; bottom: the resulting modulated structures.

ate modulated structure, in which no exact periodicity occurs, although **t** remains a pseudo-period. A modulation function can also represent a fluctuation in atomic positions and the positional modulation can also be either commensurate or incommensurate. In practice a modulated structure can simultaneously include different kinds of occupational and/or positional modulations.

A common feature of modulated structures in reciprocal space is that the reflections can be classified into two categories, one systematically strong while the other is systematically weak. This is due to the pseudo-translational symmetry. The 'strong' subset of reflections has contributions both from the 'average structure' and from the modulation functions whereas the 'weak' subsets have contributions from the modulation functions alone.

Since the contributions from the modulation functions are small by comparison with those from the average structure, a set of structure factor amplitudes corresponding to the average structure can be derived approximately from the intensities of the 'strong' reflections. Usually it is straightforward to determine the phases of the 'strong' reflections by traditional methods. However, the solution of the phase problem for the 'weak' reflections needs some special technique.

7.4.1 Direct methods for superstructures

In a superstructure there exists at least one pseudo-translation vector $\mathbf{t} = \mathbf{T}/n$, where n is an integer, $\mathbf{T}$ specifies the direction and period of the modulation wave and is the shortest lattice vector parallel to $\mathbf{t}$. If $\mathbf{r}_j$ is the average position in the pseudo-cell and f_j the scattering factor for the fraction of the atom representing average occupancy then the structure factor can be written approximately as

$$\mathbf{F}(\mathbf{h}) \simeq \sum_{j=1}^{N/n} f_j e^{2\pi i \mathbf{h} \cdot \mathbf{r}_j} \left[1 + e^{2\pi i \mathbf{h} \cdot \mathbf{t}} + e^{2\pi i \mathbf{h} \cdot 2\mathbf{t}} + \cdots + e^{2\pi i \mathbf{h} \cdot (n-1)\mathbf{t}} \right]. \tag{7.4}$$

The sum of the series inside the brackets of (7.4) is given by

$$S = \left[\exp(2\pi i \mathbf{h} \cdot n\mathbf{t}) - 1 \right] / \left[\exp(2\pi i \mathbf{h} \cdot \mathbf{t}) - 1 \right] = \begin{cases} n, & \text{if } \mathbf{h} \cdot \mathbf{t} = \text{integer} \\ 0, & \text{if } \mathbf{h} \cdot \mathbf{t} \neq \text{integer} \end{cases} \tag{7.5}$$

since $\mathbf{h} \cdot n\mathbf{t} = \mathbf{h} \cdot \mathbf{T} = $ integer. The equation is only approximate since, if the positions are modulated then $\mathbf{r}_j$ will actually vary from one pseudo-cell to the next or, if occupation is modulated then f_j will not be the same in all cells. However, it is evident that all reflections with $\mathbf{h} \cdot \mathbf{t}$ not equal to an integer will be systematically very weak, leading to an effect of pseudo-systematic extinction. The difficult part of solving a superstructure is to derive phases for the systematically weak reflections. Many attempts have been made to solve this problem by direct methods (Fan, 1975; Fan, He, Qian and Liu, 1978; Gramlich, 1975, 1978; Böhme, 1982; Prick, Beurskens and Gould, 1983 and papers in the proceedings of the *Pre-meeting Workshop on Direct Methods and their Application to Structures Showing Superstructure Effects*, 9th European Crystallographic Meeting, Torino, Italy, 1985). A very successful phase-extension technique for solving superstructures has been developed and automated in a modified version of the programme MULTAN-80 and in the programme SAPI (Fan, Yao, Main and Woolfson, 1983; Fan, Yao and Qian 1988; Fan, Qian, Zheng, Gu, Ke and Huang, 1990).

The underlying principle of the method is to consider a superstructure as the superposition of a 'difference structure' on to an average structure, the latter fulfilling exactly the pseudo-translational symmetry. This gives

$$\rho(\mathbf{r}) = \rho(\mathbf{r})_{av} + \Delta\rho(\mathbf{r}).\qquad(7.6)$$

By squaring both sides of (7.6) and neglecting the term $\Delta\rho^2(\mathbf{r})$, it follows that

$$\rho^2(\mathbf{r}) = \rho^2(\mathbf{r})_{av} + 2\rho(\mathbf{r})_{av} \cdot \Delta\rho(\mathbf{r}).\qquad(7.7)$$

From (7.7), by following the derivation of Sayre (1952) relating the structure factor of squared electron density to that of the normal density, there is obtained

$$\mathbf{F}(\mathbf{h})_{wk} = 2(\theta/V)\sum_{\mathbf{h}'}\mathbf{F}(\mathbf{h}')_{st}\mathbf{F}(\mathbf{h}-\mathbf{h}')_{wk},\qquad(7.8)$$

where $\mathbf{F}(\mathbf{h})_{wk}$ denotes the structure factor of the systematically weak reflections, $\mathbf{F}(\mathbf{h})_{st}$ denotes the structure factor of the systematically strong reflections, θ is an atomic form factor and V is the volume of the true unit cell. According to (7.8) phases of the systematically weak reflections can easily be derived based on the phases of the systematically strong ones. To do this the most efficient way is the multi-solution approach with large starting phase sets containing known phases of the *strong* reflections and randomly assigned phases for the *weak* ones. The resultant phase sets are evaluated and the correct solution is picked out by the figures of merit, which are the same as those used in MULTAN (§3.5.3).

The practical procedure used in the programme SAPI for solving superstructures contains the following steps.

(1) The entire set of reflections is divided into systematically strong and systematically weak reflection groups according to the pseudo-syste-matic extinction rule found automatically by the programme.
(2) The temperature and scaling factors are calculated for different reflection groups separately. The E-values so obtained are only used in the phase development to indicate the strength of relationships but are not used in the calculation of E-maps. For the latter purpose, E-values are calculated from the overall temperature and scaling factors of the entire set of reflections.
(3) Triple phase relationships involving three systematically weak reflec-tions are neglected in the phase development.
(4) The tangent refinement is carried out in two stages. In the first stage the phases of systematically strong reflections are developed in the usual way and all the 'weak' reflections are neglected. In the second stage

those 'strong' reflections previously developed with a weight greater than 0.9 are treated as known phases, while the systematically 'weak' reflections together with the remaining 'strong' reflections are then developed.

The above method has been tested on a number of known, as well as originally unknown, structures with pseudo-translational symmetry (Fan, Qian, Zheng, Gu, Ke and Huang, 1990; Fan, Yao, Zheng, Gu and Qian, 1991). One of the earliest examples (Fan, 1975) is given here.

SHAS, $C_5H_6O_5N_3K$, crystallises in space group $P2_12_12_1$ with unit cell parameters $a = 7.51$ Å, $b = 9.95$ Å, $c = 10.98$ Å and $Z = 4$. The arrangement of the heavy atoms (potassium) gives rise to a sub-periodicity of $\mathbf{t} = (\mathbf{a} + \mathbf{b} + \mathbf{c})/2$. Consequently the reflections with $h + k + l$ even are systematically strong, while those with $h + k + l$ odd are systematically weak. Fig. 7.5a shows the Fourier projection along the a axis calculated with the heavy-atom phases. In this projection there exist two pseudo-mirror planes shown as two dotted lines parallel to the b and c axes respectively. These planes originate from the pseudo-translational symmetry $\mathbf{t}$, which creates problems in solving the complete structure. By using equation (7.8) phases of the weak subset were easily derived, based on those of the strong subset. The resulting Fourier projection is shown in fig. 7.5b, in which the pseudosymmetry has been completely eliminated.

7.4.2 Direct methods for incommensurate modulated structures

An incommensurate modulated structure produces a three-dimensional diffraction pattern, which contains satellites round the main reflections. An example of a section of such a three-dimensional diffraction pattern is shown schematically in fig. 7.6.

The main reflections are consistent with a regular three-dimensional reciprocal lattice although the satellites do not fit the same lattice. On the other hand, although the satellites are not commensurate with the main reflections, they have their own periodicity. Hence, it can be imagined that the three-dimensional diffraction pattern is a projection of a four-dimensional reciprocal lattice, in which the main and the satellite reflections are all regularly situated at the lattice nodes. From the properties of the Fourier transform the incommensurate modulated structure here considered can be regarded as a three-dimensional 'section' of a four-dimensional periodic structure. This representation was first proposed by de Wolff (1974) to simplify the structure analysis of the incommensurate modulated structure of γ-Na_2CO_3.

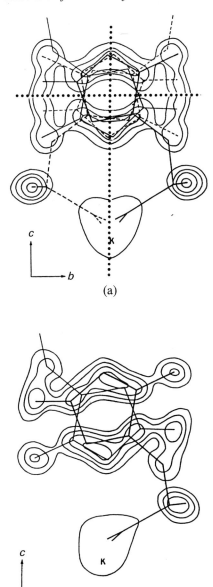

(a)

(b)

Fig. 7.5 Resolving translational ambiguity by making use of the modified Sayre equation (7.8): (a) Fourier map calculated with the heavy-atom phases; and (b) Fourier map calculated with heavy-atom phases for the 'strong' reflections and direct-method phases for the 'weak reflections'.

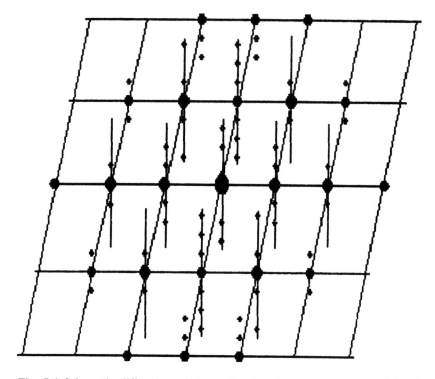

Fig. 7.6 Schematic diffraction photograph of an incommensurate modulated structure. The vertical line segments indicate the projection of lattice lines parallel to the fourth dimension.

The above example corresponds to a one-dimensional modulation. For an n-dimensional ($n = 1,2,\ldots$) modulation, a $(3+n)$-dimensional description is needed. A $(3+n)$-dimensional reciprocal vector is expressed as

$$\mathbf{h} = h_1\mathbf{b}_1 + h_2\mathbf{b}_2 + h_3\mathbf{b}_3 + \ldots + h_{3+n}\mathbf{b}_{3+n} \qquad (n = 1,2,\ldots)$$

where $\mathbf{b}_i$ is the ith translation vector defining the reciprocal unit cell. The structure factor formula is written as

$$\mathbf{F}(\mathbf{h}) = \sum_{j=1}^{N} g_j(\mathbf{h}) \exp\left[2\pi\mathrm{i}(h_1\bar{x}_{j1} + h_2\bar{x}_{j2} + h_3\bar{x}_{j3})\right], \qquad (7.9)$$

where

$$g_j(\mathbf{h}) = f_j(\mathbf{h}) \int_0^1 \mathrm{d}\bar{x}_4 \ldots \int_0^1 \mathrm{d}\bar{x}_{3+n} P_j(\bar{x}_4,\ldots,\bar{x}_{3+n})$$

$$\times \exp\{2\pi i[(h_1 U_{j1} + h_2 U_{j2} + h_3 U_{j3}) + (h_4 x_{j4} + \ldots + h_{3+n} x_{j(3+n)})]\}. \quad (7.10)$$

The $f_j(\mathbf{h})$ on the right-hand side of (7.10) is the ordinary atomic scattering factor, P_j is the occupational modulation function and U_j describes the deviation of the jth atom from its average position. For more details on (7.9) and (7.10) the reader is referred to the papers by de Wolff (1974), Yamamoto (1982) and Hao, Liu and Fan (1987). What should be emphasised here is that, according to (7.9) a modulated structure can be regarded as a set of 'modulated atoms' situated at their average positions in three-dimensional space. The 'modulated atom' in turn is defined by a 'modulated atomic scattering factor' expressed as (7.10). With this description, we have the Sayre equation in multi-dimensional space (Hao, Liu and Fan, 1987):

$$\mathbf{F}(\mathbf{h}) = (\theta/V)\sum_{\mathbf{h}'}\mathbf{F}(\mathbf{h}')\mathbf{F}(\mathbf{h}-\mathbf{h}'). \quad (7.11)$$

The right-hand side of (7.11) can be split into three parts, i.e.

$$\mathbf{F}(\mathbf{h}) = (\theta/V)\left(\sum_{\mathbf{h}'}\mathbf{F}_m(\mathbf{h}')\mathbf{F}_m(\mathbf{h}-\mathbf{h}')\right.$$
$$+ 2\sum_{\mathbf{h}'}\mathbf{F}_m(\mathbf{h}')\mathbf{F}_s(\mathbf{h}-\mathbf{h}')$$
$$\left. + \sum_{\mathbf{h}'}\mathbf{F}_s(\mathbf{h}')\mathbf{F}_s(\mathbf{h}-\mathbf{h}')\right). \quad (7.12)$$

Here subscript m stands for main reflections while subscript s stands for satellites. Since the intensities of satellites are on average much weaker than those of main reflections, the last summation on the right-hand side of (7.12) is negligible in comparison with the second, while the last two summations on the right-hand side of (7.12) are negligible in comparison with the first. Letting $\mathbf{F}(\mathbf{h})$ on the left-hand side of (7.12) represent only the structure factor of main reflections we have to a first approximation

$$\mathbf{F}_m(\mathbf{h}) \simeq (\theta/V)\sum_{\mathbf{h}'}\mathbf{F}_m(\mathbf{h}')\mathbf{F}_m(\mathbf{h}-\mathbf{h}'). \quad (7.13)$$

On the other hand, if $\mathbf{F}(\mathbf{h})$ on the left-hand side of (7.12) corresponds only to satellites, it follows that

$$\mathbf{F}_s(\mathbf{h}) \simeq 2(\theta/V)\sum_{\mathbf{h}'}\mathbf{F}_m(\mathbf{h}')\mathbf{F}_s(\mathbf{h}-\mathbf{h}'). \quad (7.14)$$

Notice that in this case the first summation on the right-hand side of (7.12) has vanished, because any three-dimensional reciprocal lattice vector corresponding to a main reflection will have zero components in the extra dimensions so that the sum of two such lattice vectors could never give rise to a lattice vector corresponding to a satellite. An exception to this can be found only when the average structure itself is a four- or higher-dimensional periodic structure as in the so-called composite structures, which are to be discussed in the following section. Equation (7.13) indicates that the phases of main reflections can be derived by a conventional direct method neglecting the satellites. Equally, equation (7.14) can be used for the phase extension from the main reflections to the satellites. This provides a way to determine directly the modulation functions. Structural details of the incommensurate modulation of the Pb-doped $Bi_2Sr_2Ca_2Cu_3O_x$ high-T_c superconductor have been observed for the first time by this method (Mo, Cheng, Fan, Li, Sha, Zheng, Li and Zhao, 1992). Since single crystals suitable for X-ray diffraction analysis are extremely difficult to prepare for this compound, electron diffraction data were used. The $0kl$ electron diffraction pattern shows satellites corresponding to a one-dimensional incommensurate modulation along the **b** axis. The crystals belong to the four-dimensional space group $P_{1\bar{1}1}^{Bbmb}$ (de Wolff, Janssen and Janner, 1981) with a three-dimensional unit cell $a = 5.49$ Å, $b = 5.41$ Å, $c = 37.1$ Å, $\alpha = \beta = \gamma = 90°$ and the modulation wave vector $\mathbf{q} = 0.117\mathbf{b}^*$. The four-dimensional real and reciprocal unit cells are defined respectively as

$$\mathbf{a}_1 = \mathbf{a}, \ \mathbf{a}_2 = \mathbf{b} - 0.117\mathbf{d}, \ \mathbf{a}_3 = \mathbf{c}, \ \mathbf{a}_4 = \mathbf{d},$$

$$\mathbf{b}_1 = \mathbf{a}^*, \ \mathbf{b}_2 = \mathbf{b}^*, \ \mathbf{b}_3 = \mathbf{c}^*, \ \mathbf{b}_4 = 0.117\mathbf{b}^* + \mathbf{d},$$

where **d** is the unit vector normal to the three-dimensional space, i.e. a unit vector simultaneously perpendicular to the vectors $\mathbf{a}, \mathbf{b}, \mathbf{c}, \mathbf{a}^*, \mathbf{b}^*$ and $\mathbf{c}^*$. The phases of $0kl0$ main reflections were calculated from the known average structure (Sequeira, Yakhmi, Iyer, Rajagopal and Sastry, 1990), while the phases of $0klm$ satellite reflections were derived by the phase extension starting from the known phases of main reflections. A Fourier map was then calculated, which is the four-dimensional potential distribution function projected along the **a** axis, i.e. $\int_0^1 \varphi(x_1, x_2, x_3, x_4) \, dx_1$, where x_1, x_2, x_3 and x_4 are fractional coordinates in the direction of $\mathbf{a}_1, \mathbf{a}_2, \mathbf{a}_3$ and $\mathbf{a}_4$ respectively. A section of $\int_0^1 \varphi(x_1, x_2, x_3, x_4) \, dx_1$ at $x_2 = 0$ is shown in fig. 7.7.

An atom without modulation will be a straight bar parallel to the fourth dimension $\mathbf{a}_4$. Occupational modulation causes a periodic variation on the width while positional modulation causes a periodic variation on the direction of the bar. It is seen that the potential at the sites of all metal atoms clearly varies along the $\mathbf{a}_4$ direction. The conclusion is that occupational modulation exists for all the metal atoms. In addition the coordinate x_3 for

0.5

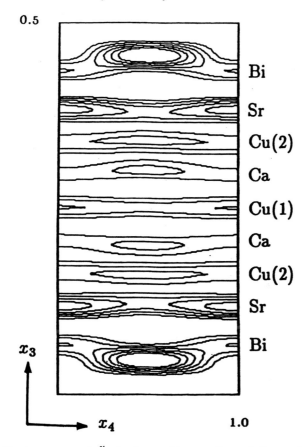

Fig. 7.7 Contour map of $\int_0^1 \varphi(x_1, 0, x_3, x_4) dx_1$, section of the projected four-dimensional potential distribution function of the Pb-doped Bi-2223 high-T_c superconductor.

the maximum potential of all metal atoms except Cu(1) also varies with x_4, indicating the existence of positional modulation for all except the Cu(1) metal atoms. The Fourier projection $\varphi(y,z)$ of the incommensurate modulated structure in the three-dimensional real space is obtained by cutting $\int_0^1 \varphi(x_1, x_2, x_3, x_4) dx_1$ normal to the $\mathbf{a}_4$ axis, where y and z are fractional coordinates in the direction of $\mathbf{b}$ and $\mathbf{c}$ axes respectively. The result is shown in fig. 7.8.

Apart from the strong modulation of metal atoms, it is seen that there are oxygen atoms disordered perpendicular to the $\mathbf{b}$ axis bridging the layers of Cu(2)–Ca–Cu(1)–Ca–Cu(2). Occupational and positional modulations of the disordered oxygen atoms are also observed.

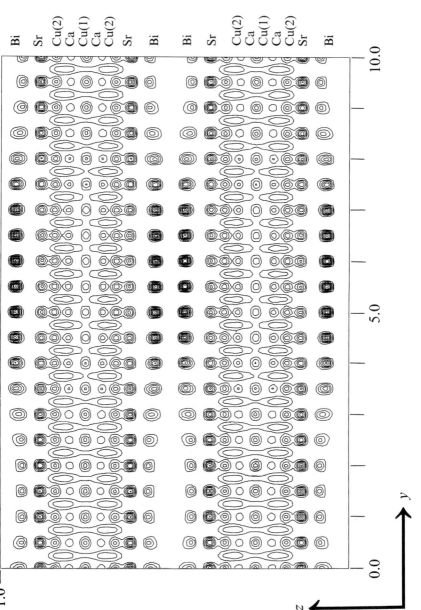

Fig. 7.8 Contour map of $\varphi(y, z)$, three-dimensional potential distribution function of Pb-doped **Bi**-2223 phase projected along the a axis. Ten unit cells are plotted along the b axis, showing the period of modulation equal to approximately 8.5 times the length of b.

7.4.3 Direct methods for composite structures

Composite structures are one kind of intergrowth structure, the characteristic of which is the coexistence of two or more mutually incommensurate three-dimensional lattices. Owing to the interaction of coexisting lattices, composite structures are also incommensurate modulated structures. The multi-dimensional Sayre equation (7.12) holds also for composite structures. Unlike ordinary incommensurate modulated structures, composite structures do not have a three-dimensional average (basic) structure. The basic structure of a composite structure can only be a four- or higher-dimensional periodic structure. Hence the first summation of the right-hand side of (7.12) will not be identical to zero even when $\mathbf{F(h)}$ on the left-hand side of (7.12) corresponds to a satellite reflection. Therefore instead of (7.14) we have

$$\mathbf{F}_s(\mathbf{h}) \simeq (\theta/V)\sum_{\mathbf{h}'}\mathbf{F}_m(\mathbf{h}')\mathbf{F}_m(\mathbf{h}-\mathbf{h}'). \tag{7.15}$$

Equation (7.15) implies that the phases of satellite reflections are uniquely determined by the phases of main reflections. This reflects exactly the fact that the modulation in composite structures comes from the interaction of different components within their basic structure. A test on applying (7.15) to determine the incommensurate modulation in composite structures $(LaS)_{1.14}NbS_2$ and $(PbS)_{1.18}TiS_2$ has been reported by Fan, van Smaalen, Lam and Beurskens (1993).

7.5 Phase extension based on electron micrographs – image processing in high resolution electron microscopy

Crystalline materials important in science and technology, such as high-T_c superconductors, are often too small in grain size and too imperfect in periodicity for an X-ray single crystal analysis to be carried out, but they are suitable for electron microscopic observation. The electron microscope is the only instrument that can produce simultaneously for a crystalline sample a micrograph and a diffraction pattern corresponding to atomic resolution. In principle, either the electron micrograph or the electron diffraction pattern could lead to a structural image.

Dorset (1991) showed that a direct-method electron diffraction analysis, based on the kinematic diffraction approximation, can be a powerful tool for crystal structure analysis. However, the phase problem in electron diffraction analysis is nothing like as easy to solve as it is in X-ray analysis.

Electron diffraction patterns provide only a partial set of three-dimensional reflections within a reciprocal sphere. This weakens the power of direct methods, since the number of phase relationships will be much decreased and some of the strongest relationships might be lost. In addition, the measurement of diffraction intensities is distorted by dynamic diffraction effects and the available techniques of intensity measurement do not compete in accuracy with those available for X-rays. This means that there are considerable difficulties in applying direct methods to electron diffraction analysis.

High-resolution electron microscopy has made great progress in recent years in the study of crystalline materials (Li, 1990). It still has two major disadvantages.

(i) In most cases a high resolution electron micrograph does not directly reveal the true structure; what is obtained is a convolution of the structural image with the Fourier transform of a contrast transfer function. Hence some technique is needed to restore the blurred image.

(ii) The point-to-point resolution of the micrograph is insufficient to resolve individual atoms in most cases. Hence some procedure is required to enhance the resolution.

The above problems can be solved by combining the information from an electron micrograph with that from the corresponding electron diffraction pattern. The procedure is in two stages. In the first stage, the electron micrograph, a blurred image of the structure at about 2 Å resolution, is deconvoluted to obtain a low-resolution structural image. By Fourier transformation of this image, a set of reliable initial phases at low resolution is obtained. In the second stage, a phase extension, using electron diffraction data, is carried out based on the initial phases obtained in the first stage. The result is a structural image at about 1 Å resolution, which is high enough to reveal individual atoms.

7.5.1 Image deconvolution

The goal of image deconvolution in high-resolution electron microscopy is to retrieve the structure image from a blurred electron micrograph, or equivalently, to extract a set of structure factors from the Fourier transform of the electron micrograph. Unlike in X-ray crystallography, where a crystal structure is represented by an electron-density function $\rho(\mathbf{r})$, a crystal structure imaged by an incident electron beam is represented by the potential distribution $\varphi(\mathbf{r})$, the Fourier transform of which yields a set of structure factors, $F(\mathbf{h})$, having the same form as (1.21) but with the atomic scattering factor replaced by that for electrons. In the following analysis we assume that the specimen under electron microscope observation is a weak-phase object – a reasonable description of any specimen consisting of light atoms with thickness less than about 100 Å. In practice the applicability of the weak-phase-object approximation has been demonstrated by Unwin and Henderson (1975) for biological specimens and by Klug (1978/1979) for an inorganic compound. Under the weak-phase-object approximation, in which dynamic diffraction effects are neglected, the Fourier transform of an electron micrograph can be expressed as

$$T(\mathbf{h}) = \delta(h) + 2\sigma F(\mathbf{h})\sin[\chi_1(h)]\exp[-\chi_2(h)], \qquad (7.16)$$

where $\delta(h)$ is a delta-function peaked at the origin, $\mathbf{h} = \mathbf{0}$, and equal to zero elsewhere. By rearrangement of (7.16), and excluding the point $\mathbf{h} = \mathbf{0}$, we obtain

$$\underset{\mathbf{h} \neq \mathbf{0}}{F(\mathbf{h})} = T(\mathbf{h})/2\sigma\sin[\chi_1(h)]\exp[-\chi_2(h)]. \qquad (7.17)$$

In (7.17) $\sigma = \pi/\lambda U$, λ is the electron wavelength, U the accelerating voltage, $\mathbf{h}$ the reciprocal vector within the resolution limit, h the magnitude of the reciprocal-lattice vector $\mathbf{h}$ and $F(\mathbf{h})$ is the structure factor of electron diffraction, which is the Fourier transform of the potential distribution, $\varphi(\mathbf{r})$, of the object. The quantity $\sin[\chi_1(h)]\exp[-\chi_2(h)]$ is the contrast transfer function, in which

$$\chi_1(h) = \pi\,\Delta f\,\lambda h^2 + \tfrac{1}{2}(\pi C_s \lambda^3 h^4),$$
$$\chi_2(h) = \tfrac{1}{2}(\pi^2\lambda^2 h^4 D^2),$$

where Δf is the defocus value, C_s is the spherical aberration coefficient and D is the standard deviation of the Gaussian distribution of defocus due to the chromatic aberration (Fijes, 1977).

The values of Δf, C_s, and D must be found for the process of image deconvolution. Of these three factors, C_s and D can be determined experimentally without much difficulty. Furthermore, in contrast to Δf, C_s and D do not change much from one image to another. This means that the main problem is the evaluation of Δf. With the estimated values of C_s and D, a set of $F(\mathbf{h})$ can be calculated from (7.17) for a given value of Δf. If the Δf value is correct then the corresponding set of $F(\mathbf{h})$ should obey the Sayre equation (3.52)

$$F(\mathbf{h}) = (\theta/V)\sum_{h'}F(\mathbf{h}')F(\mathbf{h}-\mathbf{h}'), \qquad (7.18)$$

where θ is the atomic form factor and V is the volume of the unit cell. Hence the true Δf can be found by a systematic change of the trial Δf so as to improve the validity of the Sayre equation (Han, Fan and Li, 1986). The procedure is as follows.

1. Calculate a set of $T(\mathbf{h})$ from the electron micrograph.
2. Assign trial values of Δf over a wide range with a small interval, say 10 Å. For each trial value of Δf, a set of $F(\mathbf{h})$ is calculated from $T(\mathbf{h})$ using (7.17). Reflections with $|\sin[\chi_1(h)]\exp[-\chi_2(h)]|$ less than 0.2 are neglected.
3. Calculate the figure of merit S for each set of $F(\mathbf{h})$ using the following formula (Debaerdemaeker, Tate and Woolfson, 1985):

$$S = \frac{\left[\sum_h E^*(\mathbf{h})\sum_{h'} E(\mathbf{h}')E(\mathbf{h}-\mathbf{h}')\right]^2}{\left[\sum_h |E(\mathbf{h})|^2\right]\left[\sum_h \left|\sum_{h'} E(\mathbf{h}')E(\mathbf{h}-\mathbf{h}')\right|^2\right]}, \qquad (7.19)$$

where $E(\mathbf{h})$ is the normalised structure factor and $E^*(\mathbf{h})$ is the complex conjugate of $E(\mathbf{h})$. The value of S must lie between 0 and 1 and the greater the value of S, the better does the set of $F(\mathbf{h})$ fit the Sayre equation.
4. Find the greatest S and then Fourier transform the corresponding set of $F(\mathbf{h})$ to obtain the deconvoluted image.

There are two problems with the above method.

1. The Sayre equation is used without an observed set of $|F(\mathbf{h})|$ and hence the solution for Δf may not necessarily be unique.
2. When the value of Δf is close to that of optimum defocus, the contrast transfer function is insensitive to small changes of Δf. This causes large error in the estimation of the defocus value. Both problems can be solved by introducing a set of observed $|F(\mathbf{h})|$ obtained from the corresponding electron diffraction. Then for the calculation of S, the phases derived from the electron micrograph using (7.17) and the magnitudes $|F(\mathbf{h})|$ from the electron diffraction can be combined to yield $E(\mathbf{h})$. The above procedure has been tested on a series of theoretical electron micrographs (Han, Fan and Li, 1986). Typical results are shown in fig. 7.9.

If the electron micrograph is taken near the so-called optimum defocus condition then a simple method (Liu, Xiang, Fan, Tang, Li, Pan, Uyeda and Fujiyoshi, 1990) based on Wilson statistics (Wilson, 1949) can be used

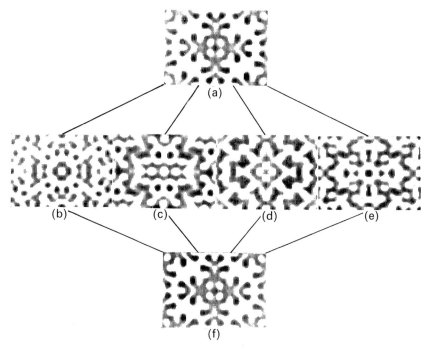

Fig. 7.9 Results on simulation of image deconvolution using a single electron micrograph and the corresponding electron diffraction pattern. (a) The expected structure image at 2 Å resolution of chlorinated copper phthalocyanine ($C_{32}N_8Cl_{16}Cu$) projected along the c axis, $a = 19.62$ Å, $b = 26.04$ Å, $c = 3.76$ Å and $\beta = 116.5°$. (b), (c), (d) and (e) The theoretical electron micrographs taken with Δf equal to -1000, $+1000$, -600 and $+600$ Å respectively, the other photographic conditions are accelerating voltage 500 kV; $C_s = 1$ mm, $D = 150$ Å. (f) The result from the deconvolution of (b), (c), (d) and (e).

for image deconvolution. Let $C(h)$ denote the contrast transfer function. Then we have from (7.17)

$$F(\mathbf{h}) = T(\mathbf{h})/[2\sigma C(h)]. \qquad (7.20)$$
$$\scriptstyle h \neq 0$$
$$\scriptstyle C(h) \neq 0$$

If the electron micrograph is taken near the optimum defocus condition, then it is possible to estimate both the magnitude and sign of $C(h)$ by a statistic approach. From (7.20), a set of $F(\mathbf{h})$ can then be obtained from the experimental data $T(\mathbf{h})$. Taking the average of $|F(\mathbf{h})|^2$ in a narrow ring with a mean radius equal to h we have according to (7.20)

$$\langle|F(\mathbf{h})|^2\rangle_h \simeq \langle|T(\mathbf{h})|^2\rangle_h/[4\sigma^2 C^2(h)]. \qquad (7.21)$$

Alternatively, we can write (Wilson, 1949)

$$\langle |F(\mathbf{h})|^2 \rangle_h \simeq \sum_{j=1}^{N} f_j^2(h), \qquad (7.22)$$

where $f_j(h)$ is the atomic scattering factor of the jth atom for X-rays and N is the number of atoms in the unit cell. By replacing $f_j(h)$ by the atomic scattering factor for electrons, (7.22) can be used for the case of electron diffraction. Combining (7.21) and (7.22) it follows that

$$C^2(h) \simeq \langle |T(\mathbf{h})|^2 \rangle_h \Big/ \left(4\sigma^2 \sum_{j=1}^{N} f_j^2(h) \right). \qquad (7.23)$$

According to (7.23) the magnitude of $C(h)$ can be obtained from experimental measurement of $T(\mathbf{h})$. It is well known that under the optimum defocus condition, $C(h)$ will be negative within a wide range of h so that, substituting (7.23) into (7.20), we finally obtain

$$F(\mathbf{h}) \simeq -T(\mathbf{h}) \left(\sum_{j=1}^{N} f_j^2(h) \right)^{1/2} \Big/ (\langle |T(\mathbf{h})|^2 \rangle_h)^{1/2}. \qquad (7.24)$$

This forms the basis of image deconvolution using a single electron micrograph under the weak-phase-object approximation and the optimum defocus condition. All that are now needed for the deconvolution are the image intensity distribution and the chemical composition of the crystalline sample. A test on an experimental electron micrograph of chlorinated copper phthalocyanine (Uyeda, Kobayashi, Ishizuka and Fujiyoshi, 1978/1979) showed that the procedure is very efficient; the result is shown in fig. 7.10.

In the process described above, during the image deconvolution, reflections with $|\sin[\chi_1(h)]\exp[-\chi_2(h)]$ smaller than a certain limit were rejected. These reflections can be retrieved by using the Sayre equation (7.18) once the phases of the other reflections are known. It is found that this considerably improves the image quality especially when the image is taken far from the optimum defocus condition.

7.5.2 *Resolution enhancement by phase extension*

An electron diffraction pattern usually contains observable reflections within a limiting sphere of 1 Å^{-1} radius. This implies that it is possible to obtain, from electron diffraction, a structural image of about 1 Å resolution, which is much higher than that obtainable from an electron micro-

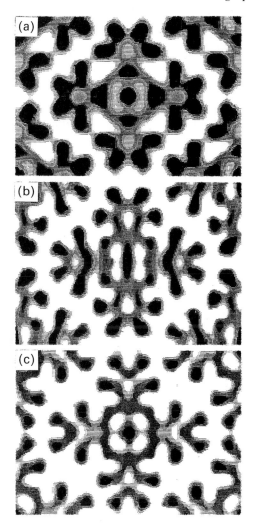

Fig. 7.10 Image deconvolution of a single experimental electron micrograph of chlorinated copper phthalocyanine: (a) the experimental image at 2 Å resolution taken near the optimum defocus condition; (b) the deconvolution result; and (c) the theoretical structure image at 2 Å resolution.

graph. In addition, the intensities of the electron diffraction pattern from a crystalline specimen are independent of defocus and the spherical aberration of the electron-optical system. Accordingly, under the weak-phase-object approximation, better quality images should be obtained from electron diffraction. However, structure analysis by electron diffraction alone is subject to the well-known difficulty of the crystallographic phase problem. On the other hand, since an electron micrograph, after deconvolution, can provide phase information corresponding to about 2 Å resolution, the complexity of the solution of the phase problem can be greatly reduced. An improved high-resolution image can be obtained by a phase extension procedure using the magnitudes of the structure factors from electron diffraction and starting phases from the corresponding electron micrograph. Phase extension techniques developed in X-ray crystallography have been used for handling electron diffraction data. The first attempt was made by Ishizuka, Miyazaki and Uyeda (1982). They applied a density modification technique to a set of error-free theoretical data of chlorinated copper phthalocyanine. Later, Fan, Zhong, Zheng and Li (1985) reported the use of a direct-method phase extension procedure to the same theoretical data. The first application to experimental data was also done on chlorinated copper phthalocyanine (Fan, Xiang, Li, Pan, Uyeda and Fujiyoshi, 1991). Fig. 7.11 shows the result of resolution enhancement of the electron micrograph of chlorinated copper phthalocyanine taken on the Kyoto 500 kV electron microscope (Uyeda, Kobayashi, Ishizuka and Fujiyoshi, 1978/1979). For comparison, results from image deconvolution of the electron micrograph and from *ab initio* direct-method phasing of the electron diffraction pattern are also given.

A similar result has also been reported for the structure of perchlorocoronene with the phase extension technique using the maximum entropy method (§7.6.7; Dong, Baird, Fryer, Gilmore, MacNicol, Bricogne, Smith, O'Keefe and Hovmöller, 1992). The direct-method phase extension technique has been further used to reveal the incommensurate modulated structure of the high-T_c superconductor Bi-2212 (Fu, Huang, Li, Li, Zhao, Cheng and Fan, 1994). The electron micrograph of Bi-2212 (Matsui and Horiuchi, 1988) is shown in fig. 7.12(a), which for the first time revealed structural features of this superconductor. Owing to the resolution of the micrograph being restricted to about 2 Å, only limited structure details can be recognised. Fig. 7.12(b) shows the corresponding electron diffraction pattern, the resolution limit of which is about 1 Å. Fourier transformation of fig. 7.12(a) followed by an image deconvolution gave the phases of a subset of main reflections in fig. 7.12(b). Starting from these reflections the direct-method phase extension enabled the determination of phases for the remaining main and satellite reflections. The Fourier map calculated with magnitudes from electron diffraction and phases from the direct method is shown in fig. 7.12(c). Individual atoms as well as details of their modulation can clearly be seen.

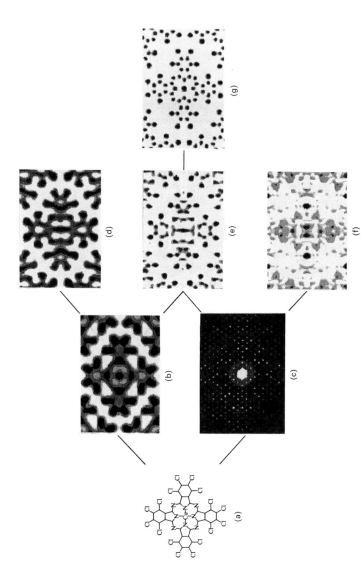

Fig. 7.11 Resolution enhancement of a single experimental electron micrograph by direct-method phase extension using the corresponding electron diffraction data and a starting phase set derived from the electron micrograph: (a) the structure model; (b) the digitised experimental electron micrograph at 2 Å resolution; (c) the corresponding electron diffraction pattern; (d) the deconvoluted experimental electron micrograph at 2 Å resolution; (e) the enhanced image obtained by combining information from (b) and (c); (f) the 'best E-map' from direct-method phasing of (c); and (g) the theoretical structure image at 1 Å.

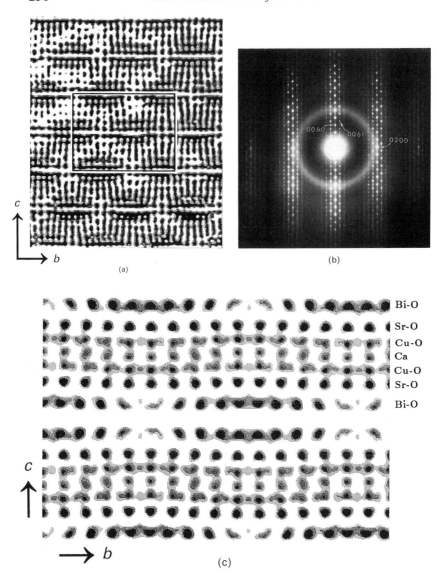

Fig. 7.12 Incommensurate modulated structure of the high-T_c superconductor Bi-2212 from the combination of electron micrograph and electron diffraction patterns: (a) the original micrograph at 2 Å resolution; (b) the corresponding electron diffraction pattern; and (c) Fourier map calculated with magnitudes from the electron diffraction pattern and phases from the direct-method phase extension based on the electron micrograph.

7.6 Phase extension for proteins

Phase extension and refinement can be achieved either by starting with a structural model or by starting with a set of initial phases. The former procedure is more reliable and accurate, if the structural model is essentially correct. However, in protein structure analysis, the starting point is often a Fourier map phased by multiple isomorphous replacement (MIR), which may not be good enough for building up a reliable structural model and it is thus important to improve the map before interpretation is attempted. To do this, phase extension and refinement, starting from a set of initial phases, is necessary.

7.6.1 Least-squares phase refinement based on the Sayre equation

The first successful phase extension and refinement process for a protein using a direct method based on the Sayre equation (3.52) was reported by Sayre (1974). Phases are adjusted to produce a minimum in the value of the expression

$$R = \sum |\alpha_h \mathbf{F}_h - \sum_{h'} \mathbf{F}_{h'} \mathbf{F}_{h-h'}|^2. \qquad (7.25)$$

Here

$$\alpha = V(p - q|\mathbf{h}| - r|\mathbf{h}|^2)(f^{sq}/f),$$

where V is the volume of the unit cell and p, q and r are parameters depending upon the degree of incompleteness of the data set. An alternative expression for α_h is

$$\alpha_h = V \sum_{h'} f_h f_{h-h'}/f_h,$$

which automatically takes into account the incompleteness of the data set. The method was applied to a protein of known structure, rubredoxin (Watenpaugh, Sieker, Herriott and Jensen, 1973). Fig. 7.13(a) shows the electron density distribution on the composite section $z = 17/60$ to $23/60$ calculated from 2.5 Å data with phases from the combination of isomorphous replacement and anomalous scattering. Direct-method phase extension resulted in a map at 1.5 Å resolution as shown in fig. 7.13(b). In comparison with the final map, fig. 7.13(c), the improvement is evident.

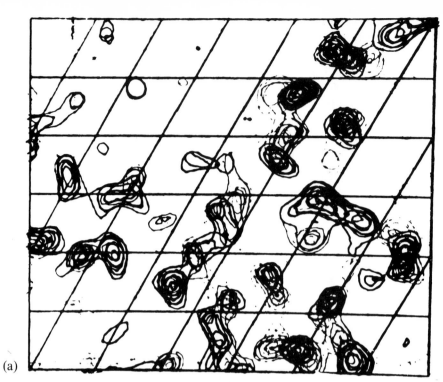

(a)

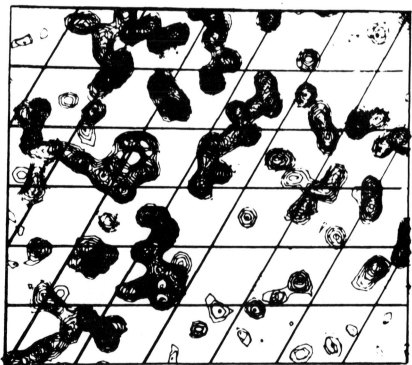

(b)

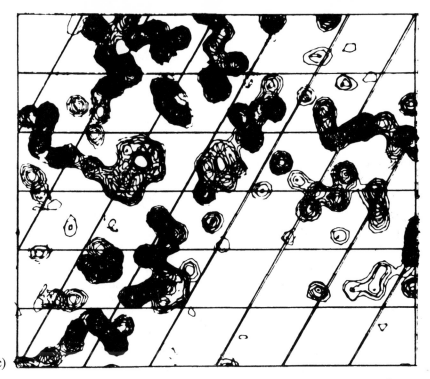

(c)

Fig. 7.13(a) Composite section $z = 17/60$ to $23/60$ for rubredoxin with 2.55 Å data and phases from a combination of isomorphous replacement and anomalous scattering; (b) direct-method phase extension to 1.5 Å; and (c) the finally-refined map.

7.6.2 Tangent formula refinement

The tangent formula (3.60) contains information corresponding to the angular portion of the Sayre equation. Many attempts have been made to use the tangent formula in phase extension and refinement for proteins but they have mostly been unsuccessful. The only exception to this is the application to the originally unknown protein avian pancreatic polypeptide by Blundell, Pitts, Tickle, Wood and Wu (1981) where phase estimates out to 2.1 Å resolution were obtained by a combination of single isomorphous replacement and one-wavelength anomalous scattering. In this application conventional structure factors rather than normalised structure factors were used in the tangent formula. In addition, a procedure was used to damp down any large changes in the phase or weight of individual reflections as the refinement proceeded. The phases of reflections within 2.1 Å resolution were extended out to 1.4 Å and the resulting electron density map could easily be interpreted.

7.6.3 Application of the Sayre-equation tangent formula

The Sayre-equation tangent formula (§3.5.5; Debaerdemaeker, Tate and Woolfson, 1985; 1988) is derived from minimisation of

$$R = \sum_h \left| E_h - (k/g_h) \sum_{h'} E_{h'} E_{h-h'} \right|^2, \tag{7.26}$$

where k is an overall scaling factor, which compensates for partial data in the $\mathbf{h}'$ summations, and g_h can be determined on a theoretical basis. In contrast to the conventional tangent formula, the Sayre-equation tangent formula relates not only the phases but also the magnitudes of structure factors. Obviously this makes it more suitable for phase extension and refinement. An example was given by Woolfson and Yao (1988) with the protein avian pancreatic polypeptide. Phases of structure factors were successfully extended from 3 Å to 1 Å with a mean phase error of only 32°.

7.6.4 The maximum determinant method

Consider a Karle–Hauptman determinant D_m (3.22) consisting of m columns and m rows. Suppose that all structure factors involved are known both in modulus and phase. Now construct a D_{m+1} determinant by adding to D_m a new column and an equivalent new row with unknown phases for some or all of the corresponding structure factors. According to the maximum determinant rule (Tsoucaris, 1970) the most probable phases for the $(m+1)^{th}$ column or row will lead to the maximum value of D_{m+1} with all eigenvalues λ_i having a positive value. Algorithms for extending phases, based on this rule, have been successfully applied to a number of known protein structures (de Rango, Mauguen and Tsoucaris, 1975a, b; Podjarny, Yonath and Traub, 1976; Podjarny and Yonath, 1977).

7.6.5 Wang's method

Apart from breaking the phase ambiguity arising from the use of single isomorphous replacement or one-wavelength anomalous scattering, the automatic solvent flattening technique (Wang, 1981, 1985) described in §5.2 has been one of the most successful methods in practice for phase extension and refinement of protein structures. The power of Wang's method is clearly seen from the crystal structure determination of troponin C from chicken skeletal muscle (Sundaralingam, Bergstrom, Strasburg, Rao, Roychowdhury, Greaser & Wang, 1985). The sample crystals belong to space group P3$_2$21 with unit cell dimensions $a = b = 66.7$ Å, $c = 60.8$ Å, $\gamma = 120°$. Phase improvement was carried out with the SIRAS (single isomorphous replacement with anomalous scattering) data based on the heavy-atom sites found with the difference Patterson maps. After eight cycles of phase improvement, the phase extension was started from 3.8 Å to 3.3 Å and then from 3.3 Å to 3.0 Å with four cycles of iteration in each stage. At the end of the process a total of 2782 reflections were phased, which is nearly double the number of 1375 starting phases at 3.8 Å. The structure was completely solved after the phase extension.

7.6.6 Histogram matching and 'SQUASH'

The density histogram of a map is the probability distribution of values at the grid points at which the map is evaluated. The histogram of a protein can be predicted without knowing its spatial structure. This property can be used in *ab initio* structure determination as has been shown by Lunin (1993). In this approach a small number (about thirty) of large reflections at low resolution (6–10 Å) are chosen and electron density maps are generated for about 5×10^5 sets of random phases. The histograms from these maps are compared with the predicted one by a χ^2 test and those of high enough likelihood are recorded. It is found that all the more likely phase sets form clusters in the multi-dimensional phase space and usually there are only two or three such clusters, one of which will be centred on the correct-phase position.

Another use of histogram matching is to improve the electron density map before constructing a structure model. Various procedures have been proposed for this purpose (Xu, 1984; Lunin, 1988; Harrison, 1988; Mariani, Luzzati and Delacroix, 1988; Zhang and Main, 1990a, b; Main, 1990a, b; Lunin and Skovoroda, 1991; Lunin and Vernoslova, 1991). The approach of Zhang and Main (1990a, b) for phase refinement and extension is considered here; it consists of the following steps.

(a) Compute the histogram of the map to be modified and acquire the expected histogram at the same resolution. The latter may be taken from the map of a similar structure or calculated from a formula (Harrison, 1988; Main, 1990b).
(b) Divide the two histograms into equal areas, as shown in fig. 7.14. This gives a list of corresponding density values ρ_i and ρ'_i, $i = 1, 2, \ldots n$ in the two histograms. A value of n of about 100 is quite satisfactory. The transformation from ρ to ρ' between the listed values is assumed to be linear.

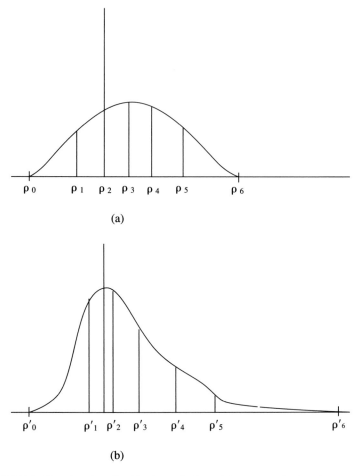

Fig. 7.14(a) The density distribution of the trial map and (b) the expected distribution. The histograms are divided into five equal areas. The transformation of density from ρ to ρ' necessary to transform the trial distribution to the expected distribution requires that ρ_i be replaced by ρ'_i.

(c) Modify the original map so that its histogram matches that of the expected one. If the value of ρ is between ρ_i and ρ_{i+1} then the transformed value, ρ', comes from

$$\frac{\rho_{i+1}-\rho}{\rho-\rho_i}=\frac{\rho'_{i+1}-\rho'}{\rho'-\rho'_i}$$

or

$$\rho'=a_i\rho+b_i, \tag{7.27}$$

where

$$a_i = \frac{\rho'_{i+1} - \rho'_i}{\rho_{i+1} - \rho_i}, \qquad b_i = \frac{\rho'_i \rho_{i+1} - \rho'_{i+1} \rho_i}{\rho_{i+1} - \rho_i}.$$

Note that this operation also applies a maximum and minimum value to the electron density, implies the correct mean value and variance and defines the entropy of the new map.

The histogram of an electron density map contains information not exploited by other techniques so far described and it seems to be particularly effective in application to phase extension and refinement. On the other hand, as pointed out by Zhang and Main (1990a), the density histogram discards all positional information. Although the histogram is unique for any particular map, vastly different maps can have identical histograms. This makes histogram matching inherently less powerful than solvent flattening since positional information is always exploited in the latter method. In view of this, histogram matching is most effective when combined with other techniques rather than being used alone.

SQUASH (Main, 1990a; Zhang and Main, 1990b) is a density modification technique combining Sayre eQUAtion, Solvent flattening and Histogram matching. Taking the inverse Fourier transform of (3.52), Sayre's equation is expressed in real space as

$$\rho(\mathbf{n}) = (V/N) \sum_m \rho^2(\mathbf{m}) \Psi(\mathbf{n} - \mathbf{m}), \tag{7.28}$$

where $\rho(\mathbf{n})$ is the electron density expressed as a discrete function evaluated at N grid points and $\Psi(\mathbf{n})$ is the inverse Fourier transform of $f(\mathbf{h})/g(\mathbf{h})$. The constraint on the electron density due to solvent flattening and histogram matching is expressed in the equation

$$\rho(\mathbf{n}) = H(\mathbf{n}), \tag{7.29}$$

where $H(\mathbf{n})$ is the electron density map modified by the combination of solvent flattening and histogram matching. Equations (7.28) and (7.29) represent a system of nonlinear simultaneous equations with as many unknowns, $\rho(\mathbf{n})$, as grid points in the asymmetric unit of the map and twice as many equations as unknowns. The least-squares solution is obtained using the Newton–Raphson technique as described by Main (1990a). This determination of the electron density forms part of the following iterative procedure of map improvement.

(a) Calculate an electron density map with approximate phases (e.g. phases from multiple isomorphous replacement (MIR)).
(b) Determine the molecular envelope as described by Wang (1981, 1985; §5.2).
(c) Set the density within the solvent region to a constant.
(d) Modify the density within the molecular envelope to match the expected histogram.
(e) Solve (7.28) and (7.29) for the electron density $\rho(\mathbf{n})$. (The map resulting from the modification in (c) and (d) is the function $H(\mathbf{n})$ in (7.29).)
(f) Calculate structure factors and their Sim weights from $\rho(\mathbf{n})$.
(g) Combine the new phases with the original phases, taking their weights into account. Extended phases and weights are accepted at their calculated values.
(h) Calculate a new map and repeat from (b) until the process has converged.

The known structure of 2Zn pig insulin was chosen as a test for the method. It was found that the combination of solvent flattening with histogram matching yields better phases than those obtained from solvent flattening alone and that the SQUASH phases are better than those obtained either by Sayre-equation refinement or by the combination of solvent flattening and histogram matching. The 3.0 Å MIR phases were refined using the SQUASH procedure and convergence was reached after five iterations. The phases were then extended to 2.0 Å in five stages, increasing the resolution by 0.2 Å at each stage. The original MIR phases at 3.0 Å, especially those of strong reflections, were significantly improved while the extended phases appeared to be just as good as those determined by MIR. The improvement in the quality of the electron density map can be seen in fig. 7.15.

SQUASH incorporates a number of different refinement criteria into a single package and it seems likely that any other competing procedure would need to have a similar characteristic.

7.6.7 The maximum entropy method

The maximum entropy method (MEM) has enjoyed much success in the field of astronomy for improving images of radio sources. The idea of applying it to crystallography originated with Narayan and Nityananda (1981, 1982) but it came into prominence following a very complete treatment given by Bricogne (1984).

Consider a unit cell divided into N pixels, and let ρ_j be the number of electrons in the jth pixel, located at $\mathbf{r}_j$. According to Shannon and Weaver (1949) and Jaynes (1957) an electron density map with minimum bias can be constructed by maximising the entropy

$$S = -\sum_j \rho_j \ln \rho_j \qquad (7.30)$$

under the constraints set by the information known with certainty. At the beginning of a structure analysis, this is a set of observed structure factors and the corresponding constraints are

$$\left| \sum_j \rho_j \exp(i2\pi\mathbf{h}\cdot\mathbf{r}_j) \right| = |F(\mathbf{h})|. \qquad (7.31)$$

In principle, the maximum entropy method can be applied to phase extension and refinement as well as to *ab initio* phase determination for proteins. However the traditional approaches, involving Lagrange multipliers, require far more computing power than even the largest computers can provide. Algorithms for simplifying the computation have been proposed (Prince, 1989; Bricogne and Gilmore, 1990) and examples of applications to known proteins have been reported (Sjölin, Prince, Svensson and Gilliland, 1991; Xiang, Carter, Bricogne and Gilmore, 1993).

Traditional direct methods involving three-phase relationships, higher order phase invariants, the tangent formula and Sayre's equation use probability formulae that are crude approximations of what can be derived from elaborate joint probability distributions obtained by the MEM approach. Thus it is fair to say that the MEM approach must be intrinsically more powerful than are traditional direct methods. However, so far it has not been found possible to exploit the full power of MEM and the need to simplify it to the point at which it can be applied has probably reduced its power somewhat. MEM concepts have been linked with many more traditional approaches – solvent flattening and Sim weights, for example, and Xiang, Carter, Bricogne and Gilmore (1993) have demonstrated with known structures that such a combined approach is very effective in phase extension and refinement. Fig. 7.16 shows the results that they obtained with the protein cytidine diaminase. With initial phases from MIRAS (Multiple Isomorphous Replacement with Anomalous Scattering) a solvent-flattened map is shown in fig.7.16a. There are breaks in the density corresponding to main-chain regions associated with an α-helix and a β-sheet. The map that comes about from the maximum entropy process is not an electron density map. However, it is possible to derive from it a

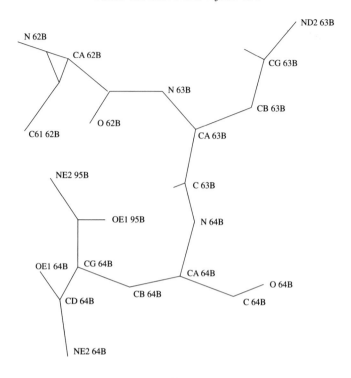

(a)

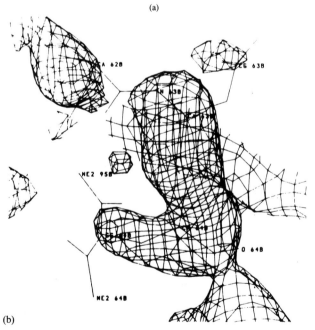

(b)

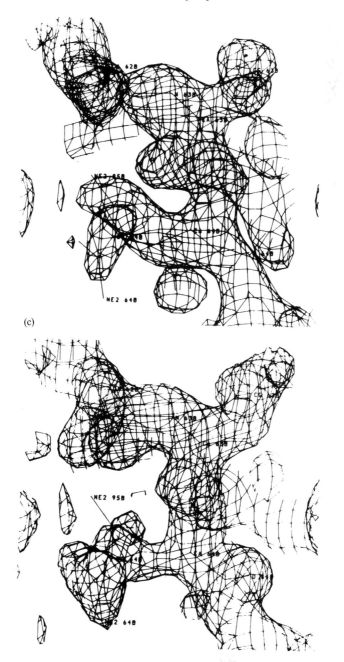

Fig. 7.15(a) Molecular structure for residues 62B to 64B of 2Zn insulin. (b) Electron density for residues 62B to 64B of 2Zn insulin obtained from the 3 Å MIR phases. (c) Same electron density as in (b) but obtained from 2 Å SQUASH phases. (d) Same electron density as in (b) but obtained at 2 Å from the refined structure.

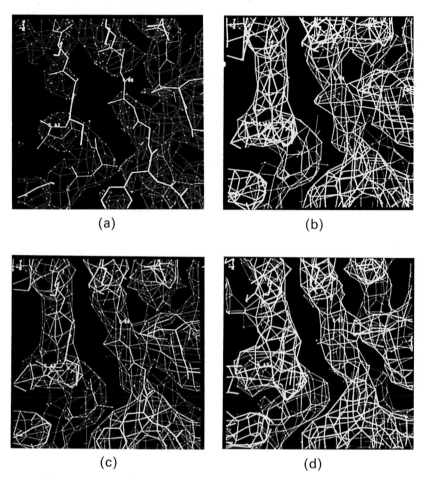

(a) (b)

(c) (d)

Fig. 7.16 Density modification of the cytidine deaminase map based on initial MIRAS phases in the region of the residues Ala 62–Cys 63 and Gly 87–Asn 89. (a) The final solvent-flattened map showing two density breaks in two segments of the main chain, in an α-helix on the left and in a β-sheet on the right. (b) A maximum-likelihood centroid map showing the least-variance electron density based on a maximum-entropy distribution. (c) A map from a combination of phase distributions from map (b) and MIRAS. (d) A $2|F_{obs}| - |F_{calc}|$ map with the finally refined phases. In addition to eliminating the breaks the process has improved the density for the side chain of Phe 86 in the lower central region of the figure.

minimum-variance estimate of the electron density, $\rho_{mv}(\mathbf{r})$, although the Fourier coefficients of $\rho_{mv}(\mathbf{r})$ will not have the observed magnitudes. This is analogous to a map produced with Sim-weighted Fourier coefficients which gives the greatest signal : noise ratio albeit that it has Fourier coefficients of the wrong magnitude. Fig. 7.16(b) shows $\rho_{mv}(\mathbf{r})$ and it will be seen that continuity is established where it was previously broken; associated with this map there are found probability distributions of the phases of individual structure factors. If the products of these probability distributions with those from MIRAS are used to obtain new phase estimates (see fig. 4.7(e)) then the map in fig. 7.16(c) is obtained. These last two maps are compared with a map using Fourier coefficients $2|F_{obs}| - |F_{calc}|$ and the refined phases, which usually gives the best representation of the structure and it is clear that the MEM process has been very effective in improving the density. With the rapid growth of computing power the limitations of MEM-based methods may eventually be overcome to the extent that they will become the dominant tool of the crystallographer.

8

Multiple-beam scattering methods

8.1 Multiple-beam scattering

8.1.1 A description of multiple-beam scattering

In fig. 1.12 there is shown a graphical means of determining the condition under which a particular diffracted beam will be produced. If the reciprocal lattice point at position **s** is the only one that touches the sphere of reflection then the associated diffracted beam will be the only one to occur – except of course for the straight-through diffracted beam, corresponding to the point O, which always occurs. Such a situation is described as two-beam diffraction. It is also possible to have more than two reciprocal lattice points on the surface of the reflecting sphere and in fig. 8.1 the points O, P and Q lie on the surface corresponding to the reciprocal-lattice vectors **0**, **h** and **k**. This would be a case of three-beam diffraction. Four- and more beam diffraction is also possible but here we shall be restricting our attention to the three-beam case.

The line CQ in fig. 8.1 gives the direction of the **k** diffracted beam. Let us suppose that a beam of radiation was incident on the crystal from that direction. The crystal would still be in the same orientation, as would be the reciprocal lattice which is rigidly attached to the crystal, and the sphere of reflection would be displaced to put the new origin point, O, on the old point Q. It is now clear that after the displacement the reciprocal lattice point **h** − **k** will now fall at the same point as P and hence that this diffracted beam would occur. In terms of the Bragg model what occurs is that the incident beam is reflected by the planes **k** to give a beam that is again reflected by the planes **h** − **k**, which then travels along a path coincident with the reflected beam **h**. If all this is taking place within one coherent block of a mosaic crystal then the resultant beam along the direction CP will depend

244

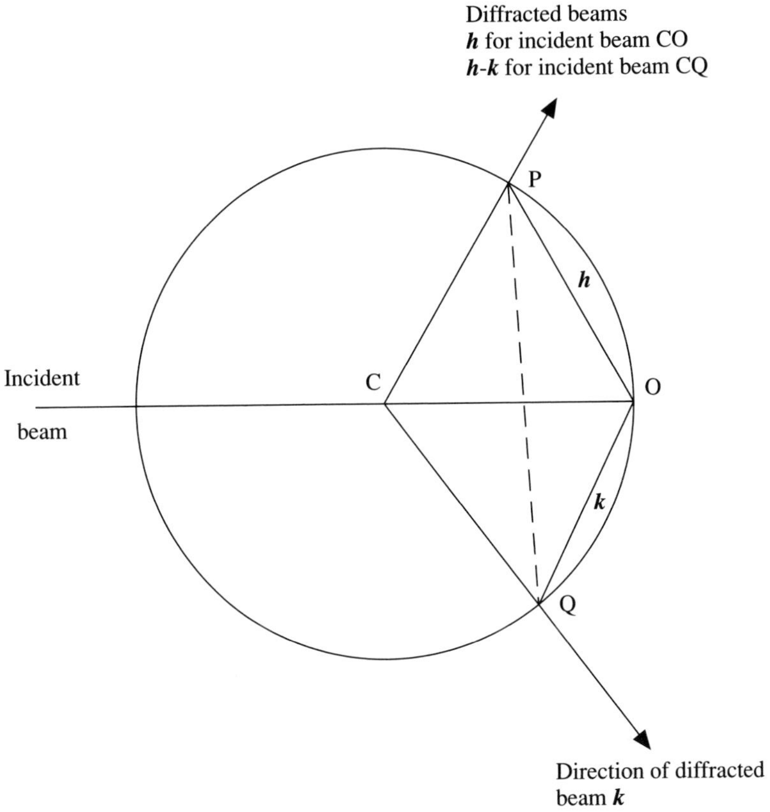

Diffracted beams
h for incident beam CO
h-k for incident beam CQ

Incident

beam

C

P

h

O

k

Q

Direction of diffracted
beam **k**

Fig. 8.1 The three-beam scattering situation. The reflected beam **k** further reflected by the lattice planes **h** − **k** along the same direction as the reflected beam **h**. Interference effects thus occur.

on interference between the components that have come along the two different paths.

The actual situation is very complex with energy being exchanged backwards and forwards between beams that are coupled together by the three-beam diffraction configuration. In the case of electron diffraction n-beam diffraction usually occurs with very large n and in addition electrons are strongly scattered by atomic potentials. The only way fully to interpret electron-diffraction patterns is by means of dynamical theory, which takes into account all the coupling processes, and this makes the deduction of structure from the diffraction pattern a difficult process (but see §7.5). Fortunately n-beam X-ray diffraction rarely occurs with large n and, indeed, if one wishes to obtain three-beam diffraction one must do careful experiments. In addition X-ray diffraction is a very weak process, only a very small proportion of the incident energy being diffracted; much more energy is usually absorbed and ends up as heat. For that reason X-ray crystallographers can usually get by with kinematic theory, which assumes no coupling between different modes of reflection. This means that the structure amplitudes that they derive directly from recorded intensities are the amplitudes of the Fourier components of the electron density in the crystal – which makes the process of structure solution simpler, if not simple.

There is one case of three-beam scattering that does occur and is noticed in normal crystallographic experimentation. In some space groups there are reflections that are systematically absent, for example, $(0, k, 0)$ reflections with k odd if there is a 2_1 axis along **b**. However, if the intensities of two reflections (h, k', l) and $(\bar{h}, k - k', \bar{l})$ are both large then successive reflections from the two sets of lattice planes can give a beam along the direction of $(0, k, 0)$, albeit that such a reflection is forbidden. These are called *Renninger reflections* and they can be a cause of confusion in trying to determine space groups. Fortunately, because they are produced by two diffraction processes they tend to be much sharper than singly diffracted beams and can usually be recognised for what they are.

Although three-beam effects are not very strong in X-ray crystallography they can be measured with precise modern equipment and they can provide information about the phases of X-ray reflections. However, before we can understand how this happens we must explore some of the phase-shift effects associated with the diffraction process.

8.1.2 The scattering phase shift

When X-radiation falls on a crystal some of it passes straight through, completely unaffected by the material of the crystal, but the remainder is scattered. We are now going to consider that part that is scattered in the forward direction. In fig. 8.2(a) we show a new wavefront produced by wavelets originating from scatterers on a previous wavefront. In fig. 8.2(b) we now consider the contribution at point P on the new wavefront from all points on the old wavefront, considered as a uniform sheet of scattering material. All the radiation from the shaded annular region will arrive at P with the same phase, lagging behind that scattered from Q by

$$\phi_r = \frac{2\pi}{\lambda}(t-d) = \frac{2\pi}{\lambda}\left[(r^2+d^2)^{\frac{1}{2}} - d\right]. \tag{8.1}$$

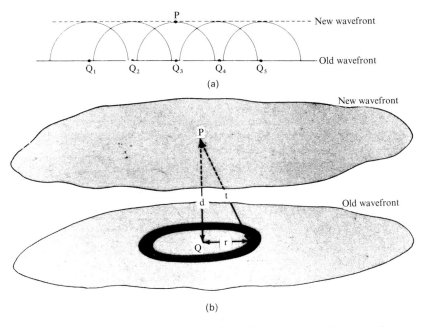

Fig. 8.2(a) The formation of a plane wavefront from wavelets originating from a plane of scatterers. (b) All radiation from the shaded annulus travels the same distance to P and arrives in phase.

The phase lag with respect to the unscattered radiation from Q is

$$\alpha_r = \frac{2\pi}{\lambda} \left[(r^2 + d^2)^{\frac{1}{2}} - d \right] + \pi, \tag{8.2}$$

where π is the Thomson scattering phase shift.

Differentiating (8.2) gives

$$\frac{d\alpha_r}{dr} = \frac{2\pi r}{\lambda (r^2 + d^2)^{1/2}}. \tag{8.3}$$

The amplitude of the contribution of the shaded annulus will be proportional to its area and inversely proportional its distance from Q, t. This will be

$$dq_r = K \frac{2\pi r \, dr}{(r^2 + d^2)^{1/2}}, \tag{8.4}$$

where K is a constant. From (8.3) and (8.4) we find, dropping the subscript r,

$$\frac{dq}{d\alpha} = K\lambda. \tag{8.5}$$

If we now construct in fig. 8.3, on an Argand diagram, the total contribution from the scattered radiation from the old wavefront at Q, with phase relative to the unscattered radiation from Q, the initial step lies along the direction AO, since, from (8.2), when $r=0$ the phase difference is π. Subsequently, from (8.5), it is seen that the length of arc is proportional to the change in phase and this describes a circle. However, scattering is strongest in the forward direction and falls with increasing scattering angle. This means that K is a slowly decreasing function of r so that the phase-vector diagram becomes a spiral rather than a circle. As r tends to infinity so the end of the spiral gradually approaches the point C in fig. 8.3, corresponding to a phase lag of $\pi/2$ of the resultant of the scattered radiation at Q relative to that which was not scattered.

This is the *scattering phase shift* and it also applies to a Bragg-reflected beam. We can now understand the process of primary extinction mentioned in §1.5.3. In fig. 8.4 a beam of radiation is moving along OA without being scattered. However, radiation reflected at C and D gives radiation moving along the same path with a phase lag of π $(\pi/2+\pi/2)$, which gives destructive interference and reduces the intensity along OA. Note that the path difference gives no phase difference because of the condition imposed by Bragg's law (1.17).

If the radiation falling on a single coherent block of a mosaic crystal has amplitude A_0 and forms a diffracted beam of index **h** then the amplitude of the diffracted beam will be

$$A(\mathbf{h}) = A_0 \alpha(\mathbf{h}) F(\mathbf{h}) e^{i\pi/2}, \tag{8.6}$$

where the final exponential term gives the scattering phase lag and $\alpha(\mathbf{h})$ is the coupling constant describing the efficiency of the scattering process. Since X-ray scattering is a weak phenomenon the values of α will generally be small.

8.1.3 Intensity change with three-beam scattering

We shall now consider the change in the intensity of the diffracted beam of index **h** in the three-beam diffraction experiment illustrated in fig. 8.1. The **h** and **k** diffraction amplitudes will be

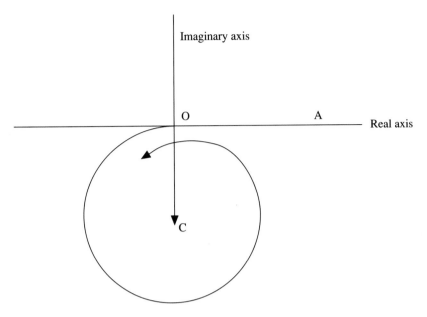

Fig. 8.3 Phase-vector diagram for radiation originating from a plane of scatterers and arriving at a plane wavefront. The resultant lies along OC.

$$A(\mathbf{h}) = A_0 \alpha(\mathbf{h})F(\mathbf{h})e^{i\pi/2}, \tag{8.6}$$

$$A(\mathbf{k}) = A_0 \alpha(\mathbf{k})F(\mathbf{k})e^{i\pi/2}. \tag{8.7}$$

The diffracted beam $\mathbf{k}$ is now diffracted again to move along the direction of diffracted beam $\mathbf{h}$ (reflected by the planes $\mathbf{h} - \mathbf{k}$) and the amplitude of the doubly diffracted beam will be

$$A(\mathbf{k}, \mathbf{h} - \mathbf{k}) = A_0 \alpha(\mathbf{k})\alpha(\mathbf{h} - \mathbf{k})F(\mathbf{k})F(\mathbf{h} - \mathbf{k})e^{i\pi}. \tag{8.8}$$

The resultant amplitude along the direction of diffracted beam $\mathbf{h}$ will be

$$A(\mathbf{h}, \mathbf{k}) = A_0 \alpha(\mathbf{h})F(\mathbf{h})e^{i\pi/2} + A_0 \alpha(\mathbf{k})\alpha(\mathbf{h} - \mathbf{k})F(\mathbf{k})F(\mathbf{h} - \mathbf{k})e^{i\pi}. \tag{8.9}$$

The ratio of the intensity of the diffracted beam $\mathbf{h}$ in the three-beam case to that in the two-beam case is found from (8.6) and (8.9) as

$$\left| \frac{A(\mathbf{h}, \mathbf{k})}{A(\mathbf{h})} \right|^2 = 1 + \left(\frac{\alpha(\mathbf{k})\alpha(\mathbf{h} - \mathbf{k})}{\alpha(\mathbf{h})} \right)^2 \left| \frac{F(\mathbf{k})F(\mathbf{h} - \mathbf{k})}{F(\mathbf{h})} \right|^2$$
$$+ 2 \frac{\alpha(\mathbf{k})\alpha(\mathbf{h} - \mathbf{k})}{\alpha(\mathbf{h})} \left| \frac{F(\mathbf{k})F(\mathbf{h} - \mathbf{k})}{F(\mathbf{h})} \right| \cos\left(\phi(\bar{\mathbf{h}}) + \phi(\mathbf{k}) + \phi(\mathbf{h} - \mathbf{k}) + \frac{\pi}{2} \right).$$
$$\tag{8.10}$$

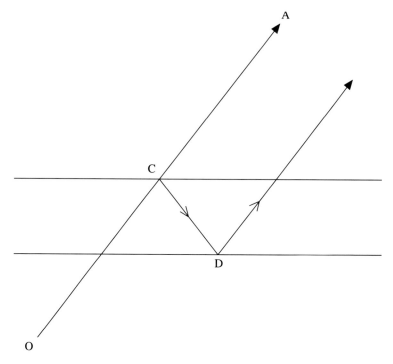

Fig. 8.4 Radiation reflected at C and D travels along the direction of the beam OA
but is π radians out of phase with it.

This expression is interesting since it contains a three-phase invariant quantity

$$\Phi_3(\mathbf{h}, \mathbf{k}) = \phi(\bar{\mathbf{h}}) + \phi(\mathbf{k}) + \phi(\mathbf{h} - \mathbf{k}), \tag{8.11}$$

so that, in principle, the measurement of the ratio of two intensities under different experimental conditions can give phase information. Given values such as

$$|F(\mathbf{h})| = 100, \ |F(\mathbf{k})| = 80, \ |F(\mathbf{h} - \mathbf{k})| = 60,$$
$$\alpha(\mathbf{h}) = \alpha(\mathbf{k}) = \alpha(\mathbf{h} - \mathbf{k}) = 2.0 \times 10^{-4}$$

the change of intensity is, at most, 2% and would be difficult, but not impossible, to measure. The experiment would involve firstly measuring the intensity in the three-beam position and then moving it to the two-beam position and recording the difference of the intensity. From the difference of intensity the cosine term in (8.10) could be evaluated, although there would still be an ambiguity in the estimated value of Φ_3.

It was shown by Post (1979) that information could be derived about the value of Φ_3, without ambiguity, by measuring the variation of intensity in the vicinity of a three-beam diffraction position. This idea was developed by Hümmer and Billy (1982, 1986) who considered the variation of intensity when the crystal was rotated about the reciprocal lattice vector **h**, so that it remained constantly in the Bragg-diffracting position while the reciprocal lattice vector **k** passed from inside to outside the sphere of reflection. The total span of the rotation considered was very small, just a few minutes of arc, so that **k** was always close to the Bragg-diffracting position. Hümmer and Billy showed that the variation of the intensity as they moved **k** from just inside to just outside the sphere of reflection depended on the value of Φ_3. In their analysis they used results from dynamical theory although a number of approximations had to be made to obtain their final result. Here we shall derive the same result using an adaptation of the kinematic approach that gave (8.10).

8.1.4 Diffraction in a near-Bragg configuration

The scenario we are going to consider is that we have a near three-beam diffraction condition. The reciprocal-lattice point **k** is just off the sphere of reflection so that the angle between the incident radiation and the reflection plane is $\theta + d\theta$, a little different from the Bragg angle θ. We are now going to consider the resultant in the direction that does make an angle θ to the reflection plane and will strike the reflection plane $\mathbf{h} - \mathbf{k}$ at the Bragg angle and so finish coincident with the diffracted beam of index **h**. The initial scattering by the planes **k** is shown in fig. 8.5. The ray OBA has travelled a distance BA − CA greater than the ray O'A in reaching the point A. From the geometry of the situation

$$\mathrm{BA} = d\operatorname{cosec}\theta,$$
$$\mathrm{CA} = \mathrm{BA}\sin\mathrm{C}\hat{\mathrm{B}}\mathrm{A} = d\operatorname{cosec}\theta\cos(2\theta + d\theta).$$

From this we find

$$\mathrm{BA} - \mathrm{CA} = 2d\sin\theta\,(1 + \cot\theta\,d\theta) = \lambda(1 + \cot\theta\,d\theta). \tag{8.12}$$

We now consider the resultant at A of all the rays scattered in the same way from planes successively more distant from A and all ending at A. The phase delay from successive planes, from (8.12), will be

$$\varepsilon = 2\pi(1 + \cot\theta\,d\theta) \equiv 2\pi\cot\theta\,d\theta \tag{8.13}$$

but in addition, because of a reduction in intensity of the beam from deeper layers due to absorption and scattering, there will be a reduction of amplitude as well. We shall indicate the factor from one layer to the next as

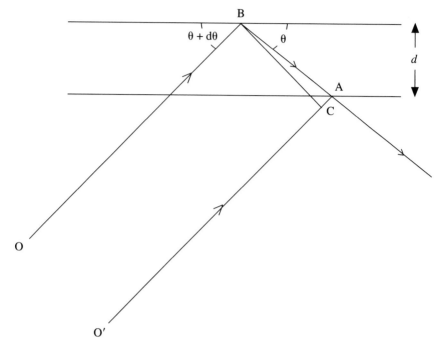

Fig. 8.5 Radiation falling on a set of reflecting planes at an angle differing by $d\theta$ from the Bragg angle θ but scattered at the Bragg angle.

η, which is very slightly less than unity. Thus the resultant disturbance at A will be

$$A_A = \sum_{j=1}^{N} \eta^j \, e^{ij\varepsilon} \qquad (8.14)$$

and this can be normalised to what it would be for the Bragg condition to give

$$R(\varepsilon) = \sum_{j=1}^{N} \eta^j \, e^{ij\varepsilon} \Bigg/ \sum_{j=1}^{N} \eta^j = |R(\varepsilon)| e^{i\xi}. \qquad (8.15)$$

Note that in (8.14) and (8.15) $e^{+ij\varepsilon}$ indicates a phase *lag* of $j\varepsilon$. The beam in the direction AD in fig. 8.6 will not be in phase with that in the same direction direction from neighbouring points. Thus in fig. 8.6 we can see that the radiation from the point A', distant x from A, will lag behind that from A by a distance

$$\text{BA}' - \text{AC} = x[\cos\theta - \cos(\theta + d\theta)] = x\sin\theta \, d\theta \qquad (8.16)$$

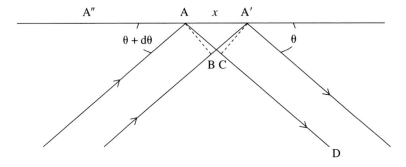

Fig. 8.6 Radiation arriving at two neighbouring points, A and A′, at an angle differing by dθ from the Bragg angle but scattered at the Bragg angle.

corresponding to a phase lag of

$$\delta = \frac{2\pi x}{\lambda} \sin\theta \, d\theta. \tag{8.17}$$

However, the beam from the point A″, the same distance on the other side of A will have a phase lag of $-\delta$ so that the sum of the disturbances from A′ and A″ will be in phase with that from A. Because incident and scattered directions are very nearly coplanar with the vector $\mathbf{k}$, the normal to the plane, this cancellation in pairs of phase lags and leads occurs to first order all over the plane and the scattering phase shift is unmodified from the Bragg condition value of $\pi/2$. Given this result the resultant amplitude of scattering, given an incident beam of amplitude A_0, is

$$A_\varepsilon(\mathbf{k}) = A_0 \alpha(\mathbf{k}) |R(\varepsilon)| F(\mathbf{k}) \, e^{i(\xi + \pi/2)}. \tag{8.18}$$

Fig. 8.7 shows the variations of $|R(\varepsilon)|$ and $\psi(\varepsilon) = \xi + \pi/2$ for N = 100 000 and $\eta = 0.9995$ for various ε. These are similar to curves given by Hümmer and Billy (1986) by an approximate dynamical-theory approach. It should be noticed that for larger positive values of ε, $\psi(\varepsilon)$ goes to zero while for larger negative values of ε, $\psi(\varepsilon)$ goes to π.

8.1.5 Profiles for ψ scans in three-beam diffraction

If the reciprocal-lattice vector $\mathbf{k}$ is inside the sphere of reflection then the incident beam will make a smaller angle to the normal to the reflecting plane than in the Bragg position and hence dθ will be positive. From (8.13) this indicates that ε will be positive; conversely with $\mathbf{k}$ outside the sphere of reflection ε will be negative. In experimental arrangements to measure the

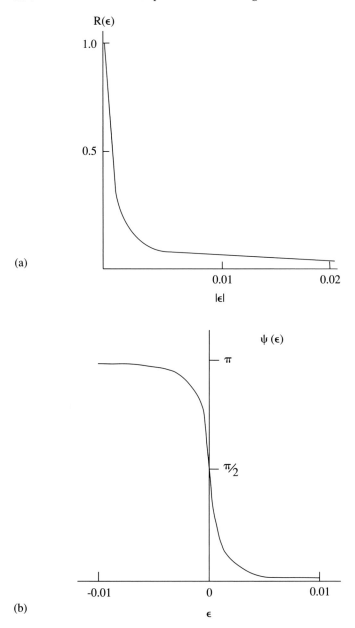

Fig. 8.7(a) $R(\varepsilon)$ as a function of ε, which describes the departure of the incident beam from the Bragg direction. $R(\varepsilon)$ is the intensity for a given ε as a fraction of that for Bragg reflection. (b) $\psi(\varepsilon)$ as a function of ε. It is $\pi/2$ for $\varepsilon=0$, 0 for large positive values of ε and π for large positive values of ε.

variation of intensity with position of $\mathbf{k}$ the crystal is rotated though a small angle about the reciprocal-lattice vector $\mathbf{h}$; this is called a ψ-scan. The convention that has been adopted is that the scan is taken from the $\mathbf{k}$-inside position to the $\mathbf{k}$-outside position so that what is measured can be compared with standard profiles.

The interference that occurs is that between the singly reflected beam of index $\mathbf{h}$ and that produced by *near-Bragg* reflection from the planes $\mathbf{k}$ and the *Bragg* reflections from the planes $\mathbf{h} - \mathbf{k}$. The scattering phase shifts from the $\mathbf{h}$ and $\mathbf{h} - \mathbf{k}$ will cancel so that the final intensity in the $\mathbf{h}$ diffraction direction is

$$|A_\varepsilon(\mathbf{h},\mathbf{k})|^2 = |A_0|^2 |\alpha(\mathbf{h})F(\mathbf{h}) + \alpha(\mathbf{k})\alpha(\mathbf{h}-\mathbf{k})|R_\varepsilon(\mathbf{k})|F(\mathbf{k})F(\mathbf{h}-\mathbf{k})e^{i\psi(\varepsilon)}|^2.$$

If this is normalised by dividing by the intensity for the two-beam configuration then the expression becomes

$$I_\varepsilon(\mathbf{h},\mathbf{k}) = \left| \frac{A_\varepsilon(\mathbf{h},\mathbf{k})}{A(\mathbf{h})} \right|^2$$

$$= 1 + \left(\frac{\alpha(\mathbf{k})\alpha(\mathbf{h}-\mathbf{k})}{\alpha(\mathbf{h})} \right)^2 |R_\varepsilon(\mathbf{k})|^2 \left| \frac{F(\mathbf{k})F(\mathbf{h}-\mathbf{k})}{F(\mathbf{h})} \right|^2$$

$$+ 2 \frac{\alpha(\mathbf{k})\alpha(\mathbf{h}-\mathbf{k})}{\alpha(\mathbf{h})} |R_\varepsilon(\mathbf{k})| \left| \frac{F(\mathbf{k})F(\mathbf{h}-\mathbf{k})}{F(\mathbf{h})} \right| \cos\left[\Phi_3(\mathbf{h},\mathbf{k}) + \psi(\varepsilon) \right]. \quad (8.19)$$

As ε goes from a positive value to a negative value ($\mathbf{k}$-inside to $\mathbf{k}$-outside) so $I_\varepsilon(h,k)$ describes a particular intensity profile. The precise form of this depends on the magnitudes of the involved structure factors and of the coupling factors α but their general form is tolerant of a large range of values of these variables. In fig. 8.8 there are shown the profiles for the same values as used in §8.3 for these variables and for $\Phi_3(\mathbf{h},\mathbf{k})$ from $0°$ to $315°$ at intervals of $45°$. It will be seen that the different profiles are quite distinctive and it should be noted that, if found experimentally, they indicate unique values of $\Phi_3(\mathbf{h},\mathbf{k})$ with no ambiguity. We shall now see what has been achieved experimentally.

8.2 Multiple-beam scattering experiments

8.2.1 Experimental determination of three-phase invariants

In order to observe the variation of X-ray intensity during a ψ-scan it is necessary rotate the crystal around a reciprocal-lattice vector. Procedures for doing this on a four-circle diffractometer have been described by several

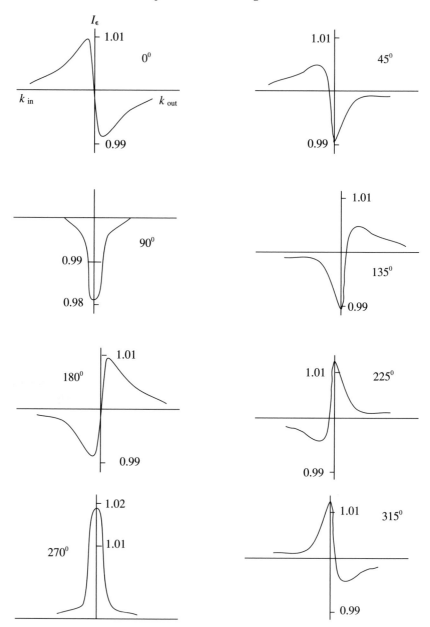

Fig. 8.8 The general form of theoretical ψ-scan profiles for various values of Φ_3.

authors, notably Han and Chang (1982). The earliest measurements of ψ-scan profiles were carried out for centrosymmetric crystals by Post (1979), Chang and Han (1982), Gong and Post (1983) and Han and Chang (1983). Since the profiles corresponding to phases of 0 and π are very distinctive the values of three-phase invariants could be measured with some degree of confidence. Han and Chang (1983) showed that the combination of direct methods with the experimentally determined values of three-phase invariants led to the straightforward solution of the structure of $Cs_{10}Ga_6Se_{14}$, space group $C2/m$, with $a = 18.233$ Å, $b = 12.889$ Å, $c = 9.668$ Å, $\beta = 108.2°$, $Z = 2$. This structure had previously defied solution by MULTAN and had been solved with great difficulty by use of the Patterson function.

Convincing measurements of ψ-scan profiles for non-centrosymmetric crystals have been reported by Hümmer, Weckert and Bondza (1989, 1990). To carry out this work they constructed a special diffractometer, with six circles, which enabled them more smoothly to rotate the crystal around a chosen reciprocal-lattice vector. All the circles were driven with computer-controlled stepper motors in steps of $0.001°$ for each axis. Such precision is necessary since the total ψ-scan for a single profile is only a few minutes of arc and if the basic reciprocal-lattice vector, $\mathbf{h}$, is not kept precisely on the axis then the error introduced can be comparable to the effect being measured. In addition the crystals used should have mosaic blocks, which either are highly parallel, which gives reinforcement of interference profiles from different blocks, or so non-parallel that the profile from one block can be targetted without interference from the others. Intermediate situations will give smearing of the measured profiles, which will make them difficult to interpret.

For their first reported measurements Hümmer, Weckert and Bondza chose the crystals of various small organic molecules, including

(i) L-asparagine monohydrate, $C_4H_8N_2O_3 \cdot H_2O$, space group $P2_12_12_1$ with $a = 5.582$ Å, $b = 9.812$ Å, $c = 11.796$ Å and $Z = 4$, and
(ii) benzil, $C_{14}H_{10}O_2$, space group 3_221 or 3_121, $a = 8.376$ Å, $b = 13.700$ Å and $Z = 3$.

One of the difficulties in this work is the large number of three-beam configurations that can occur so that rotating through $360°$ about a reciprocal-lattice vector can give several thousand, with a mean angular distance of $0.05°$ separating them. It is therefore necessary to find three-beam diffraction conditions that are free of interference from others. A second requirement is that the values of $|F(\mathbf{h})|$, $|F(\mathbf{k})|$ and $|F(\mathbf{h}-\mathbf{k})|$ should be approximately equal. This is because the coupling of the diffracted beams **h** and **k** works in both directions through the reflection **h** − **k**. If $|F(\mathbf{h})|$ and $|F(\mathbf{k})|$ are approximately equal then this two-way flow of energy, which is governed by energy-conservation principles and is phase-independent, will be roughly in balance. On the other hand, if $|F(\mathbf{k})| \ll |F(\mathbf{h})|$ then energy is lost from the reflection of index **h**; this is referred to as an *Aufhellung* effect. When $|F(\mathbf{k})| \gg |F(\mathbf{h})|$ on the other hand the intensity of the reflection of index **h** is enhanced – referred to as the *Umweganregung* effect. The two effects have the characteristic of distorting the interference profiles.

The variations of intensity that form the profile are only 1% or 2% so it is necessary to record for long enough to reach a reasonably high statistical precision. Consequently it takes several hours to measure one profile even with a rotating anode X-ray tube. Because of the likelihood that the intensity of the X-ray beam may have a gradual drift over long periods, a multiple-scan technique was employed, which means that the profile was obtained as the sum of several hundred very fast scans in which photons were counted for only 0.5 s at each step. The eventual outcome is that about 5×10^5 counts are made at each step.

In fig. 8.9(a) there is shown a profile for L-asparagine monohydrate for $|F(022)| = 17.2$, $|F(031)| = 29.1$ and $|F(0\bar{1}1)| = 23.2$, where the indices are **h**, **k** and **h** − **k** respectively. By comparison with the calculated profiles shown in fig. 8.8 it is clear that the value of Φ_3 is 180°. Fig.8.9(b) shows the profiles for the related three-beam situations with **h** = (1 $\bar{4}$ 1), **k** = (1 0 3) and **h** − **k** = (0 $\bar{4}$ $\bar{2}$) and also with $\bar{\mathbf{h}}$, $\bar{\mathbf{k}}$ and $\bar{\mathbf{h}} - \bar{\mathbf{k}}$. The values of Φ_3 must differ only in sign and their values are clearly close to 90° and − 90° respectively.

The kind of problem that may occur when the *Umweganregung* effect occurs is illustrated in fig. 8.10, which shows the profiles for benzil for **h** = (2 $\bar{1}$ 3), **k** = (0 $\bar{1}$ 3) and **h** − **k** = (2 0 0) with the values of $|F|$ equal to 13, 28 and 59 respectively and also the triplet of reflections $\bar{\mathbf{h}}$, $\bar{\mathbf{k}}$ and $\bar{\mathbf{h}} - \bar{\mathbf{k}}$. Both appear to give the characteristics of $\Phi_3 = - 90°$ although the three-phase invariants have opposite signs. It is evident that the − 90° characteristic is stronger for the first triplet, the value of which is actually − 95°. The *Umweganregung* effect can be cancelled by subtracting the second profile from the first and then the first from the second, which then gives the correct forms for approximately − 90° and + 90°.

In their later work Hümmer, Weckert and Bondza used synchrotron radiation, which offers some advantages. Most importantly, the incident beam is highly collimated, which improves angular resolution and also by choosing a suitable wavelength they were able more easily to isolate a particular three-beam diffraction situation from interference by others. When using synchrotron radiation it is necessary to take account of the high degree of polarisation of the radiation, a factor that has not been

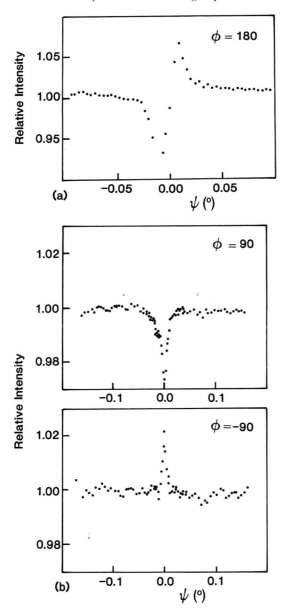

Fig. 8.9(a) The experimental ψ-span profile for a three phase-invariant of L-asparagine for $|F(\mathbf{h})| = |F(022)| = 17.2$, $|F(\mathbf{k})| = |F(031)| = 29.1$ and $|F(\mathbf{h}-\mathbf{k})| = |F(0\bar{1}1)| = 23.2$. The form of the profile shows that $\Phi_3 = 180°$. (b) Experimental ψ-scan profiles for L-asparagine for $\mathbf{h} = (1\bar{4}1)$, $\mathbf{k} = (103)$, $\mathbf{h} - \mathbf{k} = (0\bar{4}\bar{2})$ and for $\bar{\mathbf{h}}$, $\bar{\mathbf{k}}$, $\mathbf{k} - \mathbf{h}$ with $|F(\mathbf{h})| = 16.9$, $|F(\mathbf{k})| = 28.7$ and $|F(\mathbf{h}-\mathbf{k})| = 17.9$. The actual values of Φ_3 are enantiomorphically related and are indicated as 90° and −90° respectively.

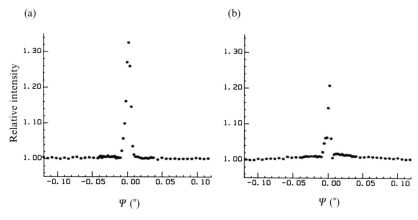

(a)   (b)

Fig. 8.10 Experimental ψ-scan profiles for benzil for $\mathbf{h} = (2\bar{1}3)$, $\mathbf{k} = (0\bar{1}3)$, $\mathbf{h} - \mathbf{k} = (200)$ and for $\bar{\mathbf{h}}$, $\bar{\mathbf{k}}$, $\mathbf{k} - \mathbf{h}$ with $|F(\mathbf{h})| = 13$, $|F(\mathbf{k})| = 28$ and $|F(\mathbf{h} - \mathbf{k})| = 59$. Because of the *Umweganregung* effect both profiles indicate about $-90°$. The actual value of Φ_3 for $\mathbf{h}$, $\mathbf{k}$ and $\mathbf{h} - \mathbf{k}$ is $-95°$.

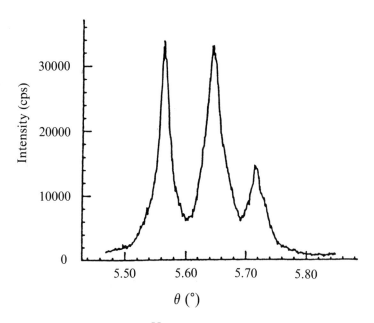

Fig. 8.11 An ω-scan through the $\bar{6}\bar{1}2$ reflection for a myoglobin crystal showing the presence of three resolvable crystalline blocks.

included in the analysis leading to (8.18). Hümmer *et al.* corrected the structure factors for the polarisation factors appropriate to the diffraction processes $(\mathbf{0}, \mathbf{h})$, $(\mathbf{0}, \mathbf{k})$ and $(\mathbf{k}, \mathbf{h} - \mathbf{k})$ where the two indices in the parentheses indicate the direction of the incident and diffracted beams.

Hümmer, Schwegle and Weckert (1991) demonstrated that triplet phases could also be estimated for a small protein; their experimental structure was that of myoglobin. For proteins a serious problem is that the reciprocal lattice is very crowded and it is difficult to avoid n-beam diffraction with n much larger than three. By controlling the wavelength of the synchrotron radiation used, they were able to find configurations for which there were only two strong reflections with reciprocal-lattice points on the sphere of reflection and the weaker ones, which inevitably were also at or near a diffracting position, had to be regarded as a source of noise on the recorded profile.

The myoglobin crystal used in the experiment contained three major mosaic blocks and a normal diffractometer ω-scan through the $(\bar{6}\ \bar{1}\ 2)$ reflection is shown in fig. 8.11. The angular separation of the blocks, about $0.1°$, was sufficient to enable a ψ-scan to be done on a single block. To optimise the conditions for good measurements the structure moduli were chosen to satisfy the condition

$$2 \leq \frac{|F'(\mathbf{k})|\,|F'(\mathbf{h} - \mathbf{k})|}{|F(\mathbf{h})|^2} \leq 6, \tag{8.20}$$

where F' is the structure factor corrected for polarisation effects. Some of the results obtained are shown in fig. 8.12. Their general quality is obviously much poorer than for the small-molecule crystals but, making intelligent allowance for *Umweganregung* and *Aufhellung* effects, it is clear that phase information can be extracted and that values of Φ_3 can be estimated with a precision of about $\pm 45°$.

8.2.2 *Making use of three-phase invariant information*

In conventional direct methods the single most important formula used is the tangent formula, (3.60). Written in the form

$$\phi(\mathbf{h}) = \text{phase of} \left(\sum_{\mathbf{k}} K(\mathbf{h}, \mathbf{k}) \exp\{i[\phi(\mathbf{k}) + \phi(\mathbf{h} - \mathbf{k})]\} \right) \tag{8.21}$$

it can be interpreted as giving an aggregated estimate of $\phi(\mathbf{h})$ where there are a number of three-phase invariants $\phi(\mathbf{h}) - \phi(\mathbf{k}) - \phi(\mathbf{h} - \mathbf{k})$, for different $\mathbf{k}$,

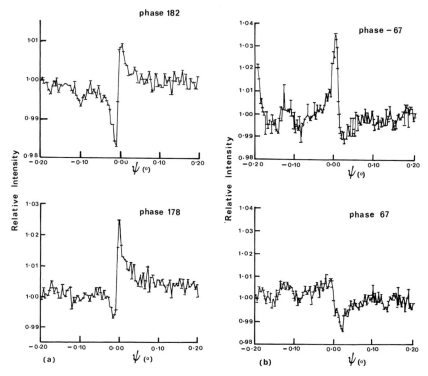

Fig. 8.12(a) Experimental ψ-scan profiles for myoglobin for $\mathbf{h} = (\bar{8}\bar{2}0)$, $\mathbf{k} = (\bar{6}01)$, $\mathbf{h} - \mathbf{k} = (\bar{2}\bar{2}\bar{1})$ and for $\bar{\mathbf{h}}, \bar{\mathbf{k}}, \mathbf{k} - \mathbf{h}$ with $|F(\mathbf{h})| = 576$, $|F(\mathbf{k})| = 1052$ and $|F(\mathbf{h} - \mathbf{k})| = 817$. Despite the noisy signal Φ_3 is indicated as 180°. The true value is 182°. (b) Experimental ψ-scan profiles for myoglobin for $\mathbf{h} = (62\bar{2})$, $\mathbf{k} = (02\bar{1})$, $\mathbf{h} - \mathbf{k} = (60\bar{1})$ and for $\bar{\mathbf{h}}, \bar{\mathbf{k}}, \mathbf{k} - \mathbf{h}$ with $|F(\mathbf{h})| = 538$, $|F(\mathbf{k})| = 1280$ and $|F(\mathbf{h} - \mathbf{k})| = 1052$. The closest fit with the profiles in fig. 8.8, consistent with the enantiomorphic relationship of the three-phase invariants, is $-45°$ and $45°$. The true values are $-67°$ and $67°$.

where each invariant has a most probable value of zero. If an estimate $\langle \phi_3(\mathbf{h}, \mathbf{k}) \rangle$, other than zero, is available then the tangent formula can be modified to

$$\tan[\phi(\mathbf{h})] = \frac{\sum\limits_{\mathbf{k}} K(\mathbf{h}, \mathbf{k}) \sin[\phi(\mathbf{k}) + \phi(\mathbf{h} - \mathbf{k}) - \langle \Phi_3(\mathbf{h}, \mathbf{k}) \rangle]}{\sum\limits_{\mathbf{k}} K(\mathbf{h}, \mathbf{k}) \cos[\phi(\mathbf{k}) + \phi(\mathbf{h} - \mathbf{k}) - \langle \Phi_3(\mathbf{h}, \mathbf{k}) \rangle]} \quad (8.22)$$

where Φ_3 is as defined in (8.11). If reliable values of $\langle \Phi_3(\mathbf{h}, \mathbf{k}) \rangle$ are available, even for a subset of three-phase invariants, then the power of the tangent formula for phase extension and refinement becomes much increased.

There is also a limitation in the usefulness of three-beam diffraction information. In theoretical direct methods the structure factor quantity that plays the main role is the normalised structure factor, large values of which can be found throughout the observed region of reciprocal space. However, three-beam diffraction experiments depend directly on having large values of $|F|$ and hence apply much more to low-resolution data. Nevertheless, despite this reservation, it is likely that n-beam diffraction, this most physical of all methods, will make a major impact on structural crystallography in the future. It could give a sufficient number of phases to define a good-quality low-resolution map amenable to phase extension by the methods described in chapter 7.

References

Baggio, R., Woolfson, M.M., Declercq, J.P. and Germain, G. (1978). *Acta Cryst.* A**34**, 883–92.

Beurskens, P.T., Bosman, W.P., Doesburg, H.M., Gould, R.O., van den Hark, Th.E.M., Prick, P.A.J., Noordik, J.H., Beurskens, G., Parthasarathi, V., Bruins-Slot, H.J., Haltiwanger, R.C., Strumpel, M.K., Smits, J.M.M., García-Granda, S., Smykalla, C., Behm, H.J.J., Schäfer, G., and Admiraal, G. (1990). *The DIRDIF-90 program system*, University of Nijmegen, The Netherlands.

Beurskens, P.T. and Prick, P.A.J. (1981). *Acta Cryst.* A**37**, 180–3.

Bhat, T.N. (1990). *Acta Cryst.* A**46**, 735–742.

Bijvoet, J.M. (1949). *Proc. Koninkl. Ned. Akad. Wetenschap.* B**52**, 313–14.

Blow, D.M. and Crick, F.H.C. (1959), *Acta Cryst.* **12**, 794–802.

Blow, D.M. and Rossmann, M.G. (1961). *Acta Cryst.* **14**, 1195–202.

Blundell, T.L., Pitts, J.E., Tickle, I.L., Wood, S.P. and Wu, C.W. (1981). *Proc. Natl. Acad. Sci. USA*, **78**, 4175–9.

Bodo, G., Dintzis, H.M., Kendrew, J.C. and Wyckoff, H.W. (1960). *Proc. Roy. Soc.* A**253**, 70–102.

Böhme, R. (1982). *Acta Cryst.* A**38**, 318–26.

Bokhoven, C., Schoone, J.C. and Bijvoet, J.M. (1951). *Acta Cryst.* **4**, 275–80.

Bragg, W.L. (1958). *Acta Cryst.* **11**, 70–5.

Bricogne, G. (1984). *Acta Cryst.* A**40**, 410–45.

Bricogne, G. and Gilmore, C.J. (1990). *Acta Cryst.* A**46**, 284–97.

Buerger, M.J. (1950). *Proc. Natl. Acad. Sci. USA* **36**, 738–42.

Buerger, M.J. (1959). *Vector Space*, New York: Wiley.

Burla, M.C., Camalli, M., Cascarano, G., Giacovazzo, C., Polidori, G., Spagna, R. and Viterbo, D. (1989). *J. Appl. Cryst.* **22**, 389–93.

Chang, S.L. and Han, F.S. (1982). *Acta Cryst.* A**38**, 414–17.

Chen, L.Q., Rose, J.P., Breslow, E., Yang, D., Chang, W.R., Furey, W.F., Sax, M. and Wang, B.C. (1991). *Proc. Natl. Acad. Sci. USA* **88**, 4240–4.

Clews, C.J.B. and Cochran, W. (1948). *Acta Cryst.* **1**, 4–11.

Cochran, W. (1952). *Acta Cryst.* **5**, 65–7.

Cochran, W. (1955). *Acta Cryst.* **8**, 473–8.

Cochran, W. and Douglas, A.S. (1955). *Proc. Roy. Soc.* A**227**, 486–500.

Cochran, W. and Woolfson, M.M. (1955). *Acta Cryst.* **8**, 1–12.

Coulter, C.L. (1965). *J. Mol. Biol.* **12**, 292–5.

Debaerdemaeker, T., Tate, C. and Woolfson, M.M. (1985). *Acta Cryst.* A**41**, 286–90.

Debaerdemaeker, T., Tate, C. and Woolfson, M.M. (1988). *Acta Cryst.* A**44**, 353–7.

Declercq, J.P., Germain, G. and Woolfson, M.M. (1975). *Acta Cryst.* A**31**, 367–72.

Dodson, G.G., Sevcik, J., Dodson, E. and Zelinka, J. (1987). *Metabolism of Nucleic Acids, Including Gene Manipulation*, pp. 33–6. Bratislava: Slovak Academy of Science.

Dong, W., Baird, T., Fryer, J.R., Gilmore, C.J., MacNicol, D.D., Bricogne, G., Smith, D.J., O'Keefe, M.A. and Hovmöller, S. (1992). *Nature* **355**, 605–9.

Dorset, D.L. (1991). *Ultramicroscopy* **38**, 23–40.

Ten Eyck, L.F. and Arnone, A. (1976). *J. Mol. Biol.* **100**, 3–11.

Fan, H.F. (1965a). *Acta Phys. Sin.* **21**, 1105–13 (in Chinese). (An English translation can be found in *Chinese Physics* (1965) 1418–28.)

Fan, H.F. (1965b). *Acta Phys. Sin.* **21**, 1114–18 (in Chinese). (An English translation can be found in *Chinese Phys.* (1965) 1429–35.)

Fan, H.F. (1975). *Acta Phys. Sin.* **24**, 57–60 (in Chinese).

Fan, H.F. and Gu, Y.X. (1985). *Acta Cryst.* A**41**, 280–4.

Fan, H.F., Han, F.S. and Qian, J.Z. (1984). *Acta Cryst.* A**40**, 495–8.

Fan, H.F., Han, F.S., Qian, J.Z. and Yao, J.X. (1984). *Acta Cryst.* A**40**, 489–95.

Fan, H.F., Hao, Q., Gu, Y.X., Qian, J.Z., Zheng, C.D. and Ke, H.M. (1990). *Acta Cryst.* A**46**, 935–9.

Fan, H.F., Hao, Q. and Woolfson, M.M. (1990a). *Acta Cryst.* A**46**, 656–9.

Fan, H.F., Hao, Q. and Woolfson, M.M. (1990b). *Acta Cryst.* A**46**, 659–64.

Fan, H.F., He L., Qian, J.Z. and Liu, S.X. (1978). *Acta Phys. Sin.* **27**, 554–8 (in Chinese).

Fan, H.F., Qian, J.Z., Zheng, C.D., Gu, Y., Ke, H.M. and Huang, S.H. (1990). *Acta Cryst.* A**46**, 99–103.

Fan, H.F., van Smaalen, S., Lam, E.J.W. and Beurskens, P.T. (1993). *Acta Cryst*, A**49**, 704–8.

Fan, H.F., Xiang, S.B., Li, F.H., Pan, Q., Uyeda, N. and Fujiyoshi, Y. (1991). *Ultramicroscopy* **36**, 361–5.

Fan, H.F., Yao, J.X., Main, P. and Woolfson, M.M. (1983). *Acta Cryst.* A**39**, 566–9.

Fan, H.F., Yao, J.X. and Qian, J.Z. (1988). *Acta Cryst.* A**44**, 688–91.

Fan, H.F., Yao, J.X., Zheng, C.D., Gu, Y.X. and Qian, J.Z. (1991). *SAPI-91, A Computer Program for Automatic Solution of Crystal Structures from X-ray Diffraction Data*, Institute of Physics, Chinese Academy of Sciences, Beijing, People's Republic of China.

Fan, H.F. and Zheng, Q.T. (1978). *Acta Phys. Sin.* **27**, 169–74 (in Chinese).

Fan, H.F., Zhong, Z.Y., Zheng, C.D. and Li, F.H. (1985). *Acta Cryst.* A**41**, 163–5.

Fijes, P.L. (1977). *Acta Cryst.* A**33**, 109–13.

Fortier, S., Moore, N.J. and Fraser, M.E. (1985). *Acta Cryst.* A**41**, 571–7.

Fu, Z.Q., Huang, D.X., Li, F.H., Li, J.Q., Zhao, Z.X., Cheng, T.Z. and Fan, H.F. (1994). *Ultramicroscopy* **54**, 229–36.

Germain, G., Main, P. and Woolfson, M.M. (1970). *Acta Cryst.* B**26**, 274–85.

Germain, G., Main, P. and Woolfson, M.M. (1971). *Acta Cryst.* A**27**, 368–76.

Germain, G. and Woolfson, M.M. (1966). *Acta Cryst.* **21**, 845–8.

Germain, G. and Woolfson, M.M. (1968). *Acta Cryst.* B**24**, 91–6.

Giacovazzo, C. (1977). *Acta Cryst.* A**33**, 933–44.
Giacovazzo, C. (1980). *Acta Cryst.* A**36**, 362–72.
Giacovazzo, C. (1983). *Acta Cryst.* A**39**, 585–92.
Giacovazzo, C. (1991). *Acta Cryst.* A**47**, 256–63.
Gillis, J. (1948). *Acta Cryst.* **1**, 76–80.
Gong, P.P. and Post, B. (1983). *Acta Cryst.* A**39**, 719–24.
Gramlich, V. (1975). X[th] International Congress of Crystallography, Amsterdam. *Acta Cryst.* A**31**, S90.
Gramlich, V. (1978). XIth International Congress of Crystallography, Warsaw. *Acta Cryst.* A**34**, S43.
Han, F.S. and Chang, S.L. (1982). *J. Appl. Cryst.* **15**, 570–1.
Han, F.S. and Chang, S.L. (1983). *Acta Cryst.* A**39**, 98–101.
Han, F.S., Fan, H.F. and Li, F.H. (1986). *Acta Cryst.* A**42**, 353–6.
Hao, Q. and Fan, H.F. (1988). *Acta Cryst.* A**44**, 379–82.
Hao, Q., Liu, Y.W. and Fan, H.F. (1987). *Acta Cryst.* A**43**, 820–4.
Hao, Q. and Woolfson, M.M. (1989). *Acta Cryst.* A**45**, 794–7.
van den Hark, Th.E.M., Prick, P.A.J. and Beurskens, P.T. (1976). *Acta Cryst.* A**32**, 816–21.
Harker, D. (1936). *J. Chem. Phys.* **4**, 381–90.
Harker, D and Kasper J.S. (1948). *Acta Cryst.* **1**, 70–5.
Harrison, R.W. (1988). *J. Appl. Cryst.* **21**, 949–52.
Hauptman, H. (1975). *Acta Cryst.* A**31**, 680–7.
Hauptman, H. (1976). *Acta Cryst.* A**32**, 934–40.
Hauptman, H. (1982a). *Acta Cryst.* A**38**, 289–94.
Hauptman, H. (1982b). *Acta Cryst.* A**38**, 632–41.
Hauptman, H. and Karle, J. (1953). *Solution of the Phase Problem. I The Centrosymmetric Crystal.* A.C.A. Monograph No.3. Willmington: The Letter Shop.
Hauptman, H. and Karle, J. (1956). *Acta Cryst.* **9**, 45–55.
Hauptman, H., Potter, S. and Weeks, C.M. (1982). *Acta Cryst.* A**38**, 294–300.
Hazell, A.C. (1970). *Nature, Lond.* **227**, 269.
Heinerman, J.J.L., Krabbendam, H., Kroon, J. and Spek, A.L. (1978). *Acta Cryst.* A**34**, 447–50.
Hendrickson, W.A. (1971). *Acta Cryst.* B**27**, 1474–5.
Hendrickson, W.A. (1985). *Trans. Am. Crystallogr. Assoc.* **21**, 11–21.
Hendrickson, W.A. and Lattman, E.E. (1970). *Acta Cryst.* B**26**, 136–43.
Hendrickson, W.A., Pähler, A., Smith, J.L., Satow, Y., Merritt, E.A. and Phizackerley, R.P. (1989). *Proc. Natl. Acad. Sci. USA* **86**, 2190–4.
Hendrickson, W.A., Smith, J.L., Phizackerley, R.P. and Merritt, E.A. (1988). *Proteins* **4**, 77–88.
Hendrickson, W.A. and Teeter, M.M. (1981). *Nature, Lond.* **290**, 107–13.
Hoppe, W. (1962). *Naturwiss.* **49**, 536.
Hoppe, W. (1963). *Z. Krist.* **118**, 121–6.
Hull, S.E. and Irwin, M.J. (1978). *Acta Cryst.* A**34**, 863–70.
Hull, S.E., Viterbo, D., Woolfson, M.M. and Zhang, Shao-hui (1981). *Acta Cryst.* A**37**, 566–72.
Hümmer, K. and Billy, H. (1982). *Acta Cryst.* A**38**, 841–8.
Hümmer, K. and Billy, H. (1986). *Acta Cryst.* A**42**, 127–33.
Hümmer, K., Schwegle, W. and Weckert, E. (1991). *Acta Cryst.* A**47**, 60–2.
Hümmer, K., Weckert, E. and Bondza, H. (1989). *Acta Cryst.* A**45**, 182–7.
Hümmer, K., Weckert, E. and Bondza, H. (1990). *Acta Cryst.* A**46**, 393–402.
Ishizuka, K., Miyazaki, M. and Uyeda, N. (1982). *Acta Cryst.* A**38**, 408–13.

Jaynes, E.T. (1957). *Phys. Rev.* **106**, 620–30.

Karle, I.L. (1974). *J. Am. Chem. Soc.* **96**, 4000–6.

Karle, I.L. (1975). *J. Am. Chem. Soc.* **97**, 4379–86.

Karle, J. (1966). *Acta Cryst.* **21**, 273–6.

Karle, J. (1968). *Acta Cryst.* **B24**, 182–6.

Karle, J. (1980). *Int. J. Quantum Chem. Symp.* **7**, 357–67.

Karle, J. (1983). *Acta Cryst.* **A39**, 800–5.

Karle, J. (1984). *Acta Cryst.* **A40**, 4–11.

Karle, J. (1989). *Acta Cryst.* **A45**, 303–7.

Karle, J. and Hauptman, H. (1950). *Acta Cryst.* **3**, 181–7.

Karle, J. and Hauptman, H. (1956). *Acta Cryst.* **9**, 635–51.

Karle, J. and Karle, I.L. (1963). *Acta Cryst.* **16**, 969–75.

Karle, J. and Karle, I.L. (1964). *Acta Cryst.* **17**, 835–41.

Karle, J. and Karle, I.L. (1966). *Acta Cryst.* **21**, 849–59.

Kartha, G. (1961). *Acta Cryst.* **14**, 680–6.

Kendrew, J.C., Dickerson, R.E., Strandberg, B.E., Hart, R.G., Davies, D.R., Phillips, D.C. and Shore, V.C. (1960). *Nature*, **185**, 422–7.

Klop, E.A., Krabbendam, H. and Kroon, J. (1987). *Acta Cryst.* **A43**, 810–20.

Klug, A. (1978/1979). *Chem. Scripta* **14**, 245–56.

Krabbendam, H. and Kroon, J. (1971). *Acta Cryst.* **A27**, 48–53.

Langs, D.A. (1986). *Acta Cryst.* **A42**, 362–8.

Lessinger, L. and Wondratschek, H. (1975). *Acta Cryst.* **A31**, 521.

Leslie, A.G.W. (1987). *Acta Cryst.* **A43**, 134–6.

Li, F.H. (1990). *Proc. International Workshop on Electron Crystallography*, Erice, Italy, pp. 153–68.

Liu, Y.W., Xiang, S.B., Fan, H.F., Tang, D., Li, F.H., Pan, Q., Uyeda, N. and Fujiyoshi, Y. (1990). *Acta Cryst.* **A46**, 459–63.

Lunin, V.Yu. (1988). *Acta Cryst.* **A44**, 144–50.

Lunin, V.Yu. (1993). *Acta Cryst.* **D49**, 90–9.

Lunin, V.Yu. and Skovoroda, T.P. (1991). *Acta Cryst.* **A47**, 45–52.

Lunin, V.Yu. and Vernoslova, E.A. (1991). *Acta Cryst.* **A47**, 238–43.

Main, P. (1990a). *Acta Cryst.* **A46**, 372–7.

Main, P. (1990b). *Acta Cryst.* **A46**, 507–9.

Mariani, P., Luzzati, V. and Delacroix, H. (1988). *J. Mol. Biol.* **204**, 165–89.

Matsui, Y. and Horiuchi, S. (1988). *Jpn. J. Appl. Phys.* **27**, L2306–9.

Metropolis, N., Rosenbluth, A.W., Rosenbluth, M.N., Teller, A.H. and Teller, E. (1953). *J. Chem. Phys.* **21**, 1087–92.

Mills, O.S. (1958). *Acta Cryst.* **11**, 620–3.

Mo, Y.D., Cheng, T.Z., Fan, H.F., Li, J.Q., Sha, B.D., Zheng, C.D., Li, F.H. and Zhao, Z.X. (1992). *Supercond. Sci. Technol.* **5**, 69–72.

Mukherjee, M. and Woolfson, M.M. (1993). *Acta Cryst.* **D49**, 9–12.

Mukherjee, A.K., Helliwell, J.R. and Main, P. (1989). *Acta Cryst.* **A45**, 715–18.

Narayan, R. and Nityananda, R. (1981). *Curr. Sci.* **50**, 168–70.

Narayan, R. and Nityananda, R. (1982). *Acta Cryst.* **A38**, 122–8.

Noordik, J.H. and Beurskens, P.T. (1971). *J. Cryst. Mol. Struct.* **1**, 339–45.

Noordik, J.H., Beurskens, P.T., Ottenheijm, H.C.J., Herscheid, J.D.M. and Tijhuis, M.W. (1978). *Cryst. Struct. Comm.* **7**, 669–77.

Nordman, C.E. (1966). *Trans. Am. Cryst. Assoc.* **2**, 29–38.

Okaya, Y., Saito, Y. and Pepinsky, R. (1955). *Phys. Rev.* **98**, 1857–8.

Okaya, Y. and Nitta, I. (1952a). *Acta Cryst.* **5**, 291.

Okaya, Y. and Nitta, I. (1952b). *Acta Cryst.* **5**, 564–70.

Okaya, Y. and Nitta, I. (1952c). *Acta Cryst.* **5**, 687.

Patterson, A.L. (1934). *Phys. Rev.* **46**, 372–6.
Pepinsky, R. (1956). *Rec. Chem. Prog.* **17**, 145–89.
Pepinsky, R. and Okaya, Y. (1957). *Phys. Rev.* **108**, 1231–2.
Pepinsky, R., Robertson, J. and Speakman, J. (editors) (1961). *Computing Methods and the Phase Problem in X-ray Crystal Analysis.* Oxford: Pergamon.
Perutz, M.F. (1956). *Acta Cryst.* **9**, 867–73.
Pitt, G.J. (1948). *Acta Cryst.* **1**, 168–74.
Podjarny, A.D. and Yonath, A. (1977). *Acta Cryst.* A**33**, 655.
Podjarny, A.D., Yonath, A. and Traub, W. (1976). *Acta Cryst.* A**32**, 281–92.
Post, B. (1979). *Acta Cryst.* A**35**, 760–3.
Prick, P.A.J., Beurskens, P.T. and Gould, R.O. (1983). *Acta Cryst.* A**39**, 570–6.
Prince, E. (1989). *Acta Cryst.* A**45**, 200–3.
Ralph, A.C. and Woolfson, M.M. (1991). *Acta Cryst.* A**47**, 533–7.
Ramachandran, J.N. and Raman, S. (1956). *Curr. Sci.* **25**, 348–51.
Ramachandran, J.N. and Raman, S. (1959). *Acta Cryst.* **12**, 957–64.
Raman, S. (1959a). *Acta Cryst.* **12**, 964–75.
Raman, S. (1959b). *Proc. Indian Acad. Sci.* A**50**, 95–107.
Rango, C. de, Mauguen, Y. and Tsoucaris, G. (1975a). *Acta Cryst.* A**31**, 227–33.
Rango, C. de, Mauguen, Y. and Tsoucaris, G. (1975b). Xth International Congress of Crystallography, Amsterdam. *Acta Cryst.* A**31**, S21.
Robbins, A.H., McRee, D.E., Williamson, M., Collett, S.A., Xuong, N.H., Furey, W.F., Wang, B.C. and Stout, C.D. (1991). *J. Mol. Biol.* **221**, 1269–93.
Sayre, D. (1952). *Acta Cryst.* **5**, 60–5.
Sayre, D. (1974). *Acta Cryst.* A**30**, 180–4.
Schenk, H. (1973a). *Acta Cryst.* A**29**, 77–82.
Schenk, H. (1973b). *Acta Cryst.* A**29**, 480–1.
Schenk, H. and De Jong, J.G.H. (1973). *Acta Cryst.* A**29**, 31–4.
Schenk, H. and Kiers, C.T. (1984). In *Methods and Applications in Crystallographic Computing*, pp. 96–105. Eds. S.R. Hall and T. Ashida, Clarendon Press, Oxford.
Sequeira, A., Yakhmi, J.V., Iyer, R.M., Rajagopal, H. and Sastry, P.V.P.S.S. (1990). *Physica* C **167**, 291–6.
Shannon, C.E. and Weaver, W. (1949). *The Mathematical Theory of Communication.* Urbana: University of Illinois Press.
Sheldrick, G.M. (1975). *SHELX. Program for Crystal Structure Determination.* University of Cambridge, England.
Sheldrick, G.M. (1990). *Acta Cryst.* A**46**, 467–73.
Sikka, S.K. (1973). *Acta Cryst.* A**29**, 211–12.
Sim, G.A. (1957). *Acta Cryst.* **10**, 536–7.
Sim, G.A. (1959). *Acta Cryst.* **12**, 813–15.
Sim, G.A. (1960). *Acta Cryst.* **13**, 511–12.
Sjölin, L., Prince, E., Svensson, L.A. and Gilliland, G.L. (1991). *Acta Cryst.* A**47**, 216–23.
Srinivasan, R. (1968). *Z. Krist.* **126**, 175–81.
Stern, F. and Beevers, C.A. (1950). *Acta Cryst.* **3**, 341–6.
Sundaralingam, M., Bergstrom, R., Strasburg, G., Rao, S.T., Roychowdhury, P., Greaser, M. and Wang, B.C. (1985). *Science* **227**, 945–8.
Toeplitz, O. (1911). *Rend. Circ. Mat. Palermo* **32**, 191–2.
Tollin, P. and Cochran, W. (1964). *Acta Cryst.* **17**, 1322–4.
Tollin, P., Main, P. and Rossman, M.G. (1966). *Acta Cryst.* **20**, 404–7.

Tsoucaris, G. (1970). *Acta Cryst.* A**26**, 492–9.

Unwin, P.N.T. and Henderson, R. (1975). *J. Mol. Biol.* **94**, 425–40.

Uyeda, N., Kobayashi, K., Ishizuka, K. and Fujiyoshi, Y. (1978/1979). *Chem. Scripta* **14**, 47–61.

Wang, B.C. (1981). *Acta Cryst.* A**37**, Suppl. C-11.

Wang, B.C. (1985). In *Methods in Enzymology*, Vol. 115: *Diffraction Methods for Biological Macromolecules.* Eds. H. Wyckoff, C.H.W. Hirs and S.N. Timasheff, New York: Academic Press. pp. 90–112.

Wang, B.C., Yoo, C.S. and Sax, M. (1979). *J. Mol. Biol.* **129**, 657–74.

Watenpaugh, K.D., Sieker, L.C., Herriott, J.R. and Jensen, L.H. (1973). *Acta Cryst.* B**29**, 943–56.

White, P. and Woolfson, M.M. (1975). *Acta Cryst.* A**31**, 53–6.

Wilson, A.J.C. (1949). *Acta Cryst.* **8**, 318–21.

Wolff, P.M.de (1974). *Acta Cryst.* A**30**, 777–85.

Wolff, P.M.de, Janssen, T. and Janner, A. (1981). *Acta Cryst.* A**37**, 625–36.

Woolfson, M.M. (1954). *Acta Cryst.* **7**, 61–7.

Woolfson, M.M. (1956). *Acta Cryst.* **9**, 804–10.

Woolfson, M.M. (1961). *Direct Methods in Crystallography.* Oxford: Oxford University Press.

Woolfson, M.M. (1970). *An Introduction to X-ray Crystallography.* Cambridge: Cambridge University Press.

Woolfson, M.M. (1977). *Acta Cryst.* A**33**, 219–25.

Woolfson M.M. and Yao, J.X. (1988). *Acta Cryst.* A**44**, 410–13.

Woolfson M.M. and Yao, J.X. (1990). *Acta Cryst.* A**46**, 9–13.

Woolfson, M.M., Yao, J.X. and Fan, H.F. (1993). *Proc. Roy. Soc.* A**442**, 13–32.

Wrinch, D.M. (1939). *Phil. Mag.* **27**, 98–122.

Xiang, S.B., Carter Jr., C.W., Bricogne, G. and Gilmore, C.J. (1994). *Acta Cryst.* D**49**, 193–212.

Xu, Z.B. (1984). *Phase extension using histogram and SIR phase information with single isomorphous replacement data* (A proposal for Ph.D. requirement), University of Pittsburgh, U.S.A.

Xu, Z.B., Yang, D.S.C., Furey, W. Jr., Sax, M., Rose, J. and Wang, B.C. (1984). *Proc. Am. Crystallogr. Assoc. Meet.* 20–25 May 1984. Lexington, Kentucky, U.S.A. Abstr. PC 2.

Yamamoto, A. (1982). *Acta Cryst.* A**38**, 87–92.

Yao, J.X. (1981). *Acta Cryst.* A**37**, 642–4.

Yao, J.X. (1983). *Acta Cryst.* A**39**, 35–7.

Yao, J.X. and Fan, H.F. (1985). *Acta Cryst.* A**41**, 284–5.

Zachariasen, W.H. (1952). *Acta Cryst.* **1**, 68–73.

Zhang, K.Y.J. and Main, P. (1990a). *Acta Cryst.* A**46**, 41–6.

Zhang, K.Y.J. and Main, P. (1990b). *Acta Cryst.* A**46**, 377–81.

Zhang, Shao-hui and Woolfson, M.M. (1982). *Acta Cryst.* A**38**, 683–5.

Sources of non-original figures

Abbreviations

IXRC   *An Introduction to X-ray Crystallography*, CUP
I.U.Cr   International Union of Crystallography

Fig. 1.14   Fig. 3.16 IXRC
Fig. 1.15   Fig. 5.29 IXRC
Fig. 1.17   Fig. 6.9 IXRC
Fig. 1.18   Fig. 4.17 IXRC
Figs. 1.19–1.24   From *International Tables for Crystallography*, I.U.Cr.

Fig. 2.3   Fig. 8.17 IXRC
Fig. 2.4   Fig. 8.15 IXRC
Fig. 2.6   Figs. 1 and 2 Stern, F. and Beevers, C.A. (1950). *Acta Cryst.* **3**, 341–6. I.U.Cr.
Fig. 2.7   Fig. 8.20 IXRC
Fig. 2.8   Fig. 8.21 IXRC

Fig. 3.5   Figs. III.2 and III.3 Woolfson, M.M. *Direct Methods in Crystallography*, OUP.

Fig. 4.8   Fig. 3 Blow, D.M. and Rossman, M.G. (1961). *Acta Cryst.* **14**, 1195–202. I.U.Cr.
Fig. 4.10   Fig. 17 Bodo, G., Dintzis, H.M., Kendrew, J.C. and Wyckoff, H.W. *Proc. Roy. Soc.* A**235**, 70–102.
Fig. 4.11   Fig. 23 *Ibid.*
Fig. 4.12   Fig. 6.17 IXRC

Fig. 6.4   Fig. 1 Fan, H.F., Hao, Q., Gu, Y.X., Qian, J.Z., Zheng, C.D. and Ke, H.M. (1990). *Acta Cryst.* A**46**, 935–9. I.U.Cr.
Fig. 6.5   Fig. 2 *Ibid.*
Fig. 6.7   Fig. 1 Hao, Q. and Woolfson, M.M. (1989). *Acta Cryst.* A**45**, 794–7. I.U.Cr.
Fig. 6.8   Fig. 1 Fan, H.F., Hao, Q. and Woolfson, M.M. (1990). *Acta Cryst.* A**46**, 656–9. I.U.Cr.

Fig. 6.9	Fig. 2 Fan, H.F., Hao, Q. and Woolfson, M.M. (1990). *Acta Cryst.* **A46**, 659–64. I.U.Cr.
Fig. 6. 11	Fig. 2 (Part A) Hendrickson, W.A., Pahler, A., Smith, J.L., Satow, Y., Merrit, E.A. and Phizackerley, R.P. (1989). *Proc. Natl. Acad. Sci.* **86**, 2190–4.
Fig. 6.14	Fig. 6 Woolfson, M.M., Yao, J.X. and Fan, H.F. (1993). *Proc. Roy. Soc.* A (in press).
Fig. 7.1	Fig. 2 Noordik, J.H. and Beurskens, P.T. (1971). *J. Cryst. Mol. Struct.* **1**, 339–45. Plenum.
Fig. 7.9	Paper by Fan, H.F. on *New developments in direct methods* in *Molecular Structure: Chemical Reactivity and Biological Activity*, p. 566, Oxford University Press.
Fig. 7.10	Fig. 2 Liu, Xiang, Fan, Tang, Li, Pan, Uyeda and Fujiyoshi (1990). *Acta Cryst.* **A46**, 459–63. I.U.Cr.
Fig. 7.12(a)	Fig. 2a Matsui, Y. and Horuichi, S. (1988). *Jpn. J. Appl. Phys.* **27**, L2306–9. Japanese Journal of Applied Physics.
Fig. 7.12(b) and (c)	Fig. 1 and Fig. 3a Fu, Z.Q., Huang, D.X., Li, F.H., Li, J.Q., Zhao, Z.X., Cheng, T.Z. and Fan, H.F. (1994). *Ultramicroscopy* **54**, 229–36.
Fig. 7.13	Fig. 4 Sayre, D. (1974). *Acta Cryst.* **A30**, 180–4. I.U.Cr.
Fig. 7.15(a)–(d)	Figs. 1–4 Zhang, K.Y.J. and Main, P. (1990). *Acta Cryst.* **A46**, 377–81. I.U.Cr.
Fig. 7.16	Fig. 7 Xiang, S., Carter, C.W., Bricigne, G. and Gilmore, C. (1993). *Acta Cryst.* **D49**, 193–212. I.U.Cr.
Fig. 8.2	Fig. 6.12 IXRC
Fig. 8.9 (a) and (b)	Part of Fig. 2 and Fig. 3 Hummer, K., Weckert, E. and Bondza, H. (1989). *Acta Cryst.* **A45**, 182–7. I.U.Cr.
Fig. 8.10	Fig. 4(c) Hummer, K., Weckert, E. and Bondza, H. (1990). *Acta Cryst.* **A46**, 393–402. I.U.Cr.
Fig. 8.11	Fig. 1 Hummer, K., Schwegle, W. and Weckert, E. (1991). *Acta Cryst.* **A47**, 60–2. I.U.Cr.
Fig. 8.12(a)	Fig. 2 *Ibid.*
Fig. 8.12(b)	Fig. 4 *Ibid.*

Index

272